Intuitive Analog to Digital Control Loops in Switchers

Part 1

by Sanjaya Maniktala

ISBN-13: **978-1518782121**

ISBN-10: **1518782124**

PREFACE

Over the years I started realizing that my chapters on control loop theory as published in both editions of Switching Power Supplies A-Z and Switching Power Supply Design and Optimization, were becoming increasingly popular and widely-referenced. I guess they were liked as they were perceived as being relatively low on math; emphasizing intuition. At least that's what a lot of blogs and reviews seemed to be saying. But it was still just a warm and fuzzy feeling on my part perhaps, as an excited author. However, I soon received dramatic confirmation of that.

But before I proceed further with that anecdote, I must point out that "natural intuition" is actually misplaced in feedback systems, since our interaction with the world around us is phase-*insensitive*. So it does take a fair amount of *math* to build up a certain form of "acquired intuition" for control loop theory, which we can then rely upon to understand far more. So, I am going to continue the development of that into the next part of this two-part series on control theory as applied to switchers. Sorry, I *will* need to delve into some math, but very judiciously I assure you. Math is not my preferred tool for self-gratification! I don't revel in personally deriving thousands of equations just because I just don't trust others' capabilities, and then end up making a ton of new mistakes of my own. All I do is enter all the abounding equations into my preferred tool, Mathcad, where I then plot them out, and follow up with a bunch of sanity checks, including for self-consistency. In no time I discover the undiscovered mistakes or typos in related literature, and simply fix them! It is faster and more efficient too. Besides, my intent has never been to intimidate my readers. I measure my success as a teacher, by how *you* benefit from it ultimately. Not how knowledgeable I look to you after the smoke has subsided.

So, coming back to my earlier story, on Jan 23, 2013, something rather interesting happened. I received a note from the well-known Dr Ray Ridley, living somewhere in France at the time. He said, "I'd like to send you a free copy of Power 4-5-6 to get your feedback on it. I would very much value your opinion on how useful it is to design engineers, and what else it might need moving forward. Please let me know if you would be able to take some time to push a few buttons and comment on it". I felt privileged. But of course Dr Ridley was only expecting me to make a few pleasant remarks on the GUI of his Excel-based software Power 4-5-6. Perhaps have me validate it and help him promote the tool with my "testimonial". However I got a bit selfish. I was in the process of writing my sixth book, the second edition of my very first book. I thought I could use the free license to double check my equations. So I went on the first screen of Ridley's Power 4-5-6 worksheet and entered the desired locations of the poles and zeros — as per the usual established practice of putting two zeros where the LC double pole was, a pole where the ESR zero was and so on. It churned out the values of the five compensation components. I then compared them with the values I got from my equations and three component values matched perfectly, ***one almost***, but ***one did not at all***. By a factor of 2, compared to my book's equations! You could argue that was no biggie. But actually, since everything is logarithmic, in the worst case it could change even the crossover frequency by a factor of 2. Further, if you look closely, each compensation component is part of one pole and also one zero. That implied *half* of the desired pole-zero locations were wrong if only one component calculation was wrong.

I was initially sure it must be my mistake. After all, this was Dr Ridley. But then there was another screen on Power 4-5-6 which said in effect: plug in the calculated component values provided by the worksheet and confirm the locations of poles and zeros. That is due diligence of course. Sure enough, using that screen I could see the error. The final poles and zeros did not match the inputted poles and zeros. Nothing to do with my book! All steps wholly contained in Power 4-5-6. But if I used the equations in my book and fixed the two wrong component values, the "self-check" panel of Power 4-5-6 turned out right!

To cut a long story short, I wrote back my findings to Dr Ridley on Feb 1, 2013, and the very next day he replied: "I think I will change Power 4-5-6….**25 years and no one ever noticed this.** I think I just lost my weekend! Do you have your schematics for a type II calculation as well?"

I had to send him those and reassure him that to err is human. On Feb 10, 2013 he released the corrected version of Power 4-5-6 based on the equations of my first-edition A-Z book…..and though he acknowledged my contribution, it was indirect. Something to the effect that he was merely trying to bring his software to be "consistent" with my book, for it was easier for him to change his software than for me to change my already published book! Well, why would he need to be consistent with me?

I was quite amazed that this major error in a paid software had been uncovered 25 years later, and *by me*. It had gone unnoticed by thousands of trusting customers who not only attended Dr Ridley's expensive seminars and workshops, but bought his consultancy services too. At the end, I just gave Dr Ridley my equations for free of course. All it took from me was an hour or two of my time on a weekend, with my very first "red book" by my side. Next to my licky maltipoo.

But there is far more coming here in this two-part book series, as you will see, when I ramp up into the second part. I will finally reveal a powerful new digital technique to dramatically improve transient response. I will also provide a completely unique way of explaining the underlying "PID" coefficients of control loop theory to you. You will understand it is far easier to visualize and manipulate than you ever thought, if you just learn to look at it "that way".

Thank you once again for your immense support over the years and your interest in this book.

Sanjaya Maniktala, October 2015

ACKNOWLEDGMENTS

I need to thank one avid reader of mine in particular, for bringing to my attention quite a few typos, and perhaps a few errors too in my previous books. He is Nicola Rosano from Italy.

I have also enjoyed great support from loyal readers over the years, across the world, but let me mention at least a few names here: Swaraj Kali, Navroop Singh, John Lee, Ajit Narwal, Malhar Bhatt, Ashish Deshpande, Amit Tiwari, Ken Coffman, Robert Gendron, Achim Döbler and Karan Goel. Thanks guys, always!

II must always remember that none of this may have been possible, if I had not met Doctor GT Murthy decades ago in Bombay.

I must also thank my wife Disha Maniktala, for standing by me always, and just letting me do what I do best (besides stopping me, often unsuccessfully, when I try to do what I don't do well!).

CONTENTS

INTRODUCTION

Scouring the enticing landscape of control loop theory for enlightenment, we can often stumble and lose our way in seemingly contradictory statements such as: "high gain *reduces* the effect of disturbances; therefore *increase* the gain". In the very same breath: "high gain causes *oscillations*; so *reduce* the gain".

Bewildering!

Here's another example: "high gain improves load regulation; so *increase* the gain". However, "*voltage positioning* improves load regulation, and for that we need to *reduce* the gain".

"Oh, by the way, don't forget to set the phase margin to *greater than 45°*. Worst-case 50°. See http://www.ridleyengineering.com/loop-stability-requirements.html. And remember, phase margin *really* just needs to be greater than *zero* to avoid oscillations. In fact, *smaller* the phase margin, snappier the response. So reduce the phase margin".

And just when you think you may have at least got the gist of it all, suddenly, you have someone advocating 75° or 76° phase margin— consistent with a "Q of 0.5". Where did Q suddenly come into the picture, by the way? See http://powerelectronics.com/power-electronics-systems/transient-response-counts-when-choosing-phase-margin and http://www.ele.uri.edu/~daly/535/margin.html . It seems futile to ask: what exactly is the relationship between "Q" (quality factor) and phase margin? Not to mention the relationship between the actual measured overshoot, to the "Q", or to the phase margin. Is the prevailing, often widely differing industry guidance regarding optimum phase margin, or Q, just based on simulations? Maybe even lab results, albeit under unknown, uncontrolled, and widely varying conditions.

We can also ask: how exactly does this optimum phase angle recommendation change in going from voltage-mode control (VMC) to current mode control (CMC)? How does it depend on the selected L and C_{OUT} of the switcher? Or on its mode of operation: continuous conduction mode (CCM) and discontinuous conduction mode (DCM)? And what is the overall topology-dependency? In other words, what happens if we are dealing with a boost or buck-boost, instead of a buck? Is the "recommended" phase margin always 45°? Or is it 76°? And why?

That is the rather confusing landscape we all encounter as we take our first baby steps into the mystical world of control loop theory.

That's when we meet the "experts". This is what they seem to be saying: The gain of the closed-loop system is called the "closed-loop gain". But it is the "open-loop gain" which we are really interested in, for that is what predicts the stability of a closed-loop system. So we do a Bode plot measurement—which incidentally, is a plot of the open-loop gain. Yes—taken on a system while its loop is closed, not open.

And so on..... How are we ever going to get to the bottom of the seeming quagmire called control loop theory?

The truth is: control loop theory is a challenging subject indeed. Even seasoned hardware engineers, so far accustomed to dealing mainly with tangible objects on a lab bench, are expected to grab a paper and pencil and start playing brainy physicist instead! They must hit the ground (plane) running, literally, leaping effortlessly through hoops of imaginary p-planes, s-planes and z-planes, and switching seamlessly between time and frequency domains. If that were not enough, they also need to embrace the fact that the frequencies involved now can not only be negative but imaginary—whatever *that* means! All this can cause the last vestiges of any old-school physical intuition to implode on itself.

It only gets worse—when we attempt to apply generic control loop theory to switchers without fully recognizing the fact that many of the concepts we struggled to absorb from all the excellent articles and papers on the subject, need serious reevaluation now. One reason for that is switchers are discrete/digital devices, not continuous as we may have assumed. That is because they have a *discrete* "control effort" update interval, related to the discrete pulses coming in at the rate of the switching frequency. The net result, expressed intuitively, is that "error" information is not necessarily sampled and communicated *instantaneously* to be acted upon and corrected—it is inherently delayed. That has major consequences. One of which is we need to lower the bandwidth of the control loop *to at least one-fifth the switching frequency*. Otherwise, we would naturally have preferred as high a bandwidth as possible. Why not?

Traditionally, a control loop system is often explained with reference to a mundane room air-conditioning system. In that example, the setting of the thermostat is the "set point", or the input (to the control loop system) — referred to as the "IN" node. The output of this closed-loop system, the "OUT" node, is the temperature of the room. A thermocouple, or sensor, is present somewhere, to monitor the room temperature. An error stage looks at the error, i.e. the difference between the set point and the output. The system incorporates negative feedback as a means of correction, so if the room temperature goes above the set point (room too "hot"), *cold* air gets pumped into the room to try to reduce the error. And so on.

We are interested in things like: what is the rate at which cold air gets pumped into the room if the error is say 15°C. And what if the error drops to 5°C? Does the rate of cold air being pumped fall proportionally—i.e. three times? Or does it just keep going at the same rate, and then simply turn off the moment it "thinks" the error has dropped to zero?

Keep in mind there could be significant delays involved in the sensing and the response—perhaps based on the non-zero specific heat capacity of the various parts constituting the sensor and blower, not to mention the objects in the room. In that case the room temperature, i.e. the output, could easily undershoot, and go below the set point. At some stage, the system will probably respond by pushing in hot air instead of cold. That too could cause an overshoot, but hopefully a diminishing error over time, so ultimately the room temperature will stabilize at the set point.

But it could do so with a quantifiable "setting error". There may be a residual 1 or 2°C error at the end of it all. Unless it can detect that error, and try to correct it. In other words: *if the system has a high "gain"*. Gain is simply Δ(OUT) divided by Δ(IN). So there will be a certain Δ(OUT) based on a finite, not infinite, gain, coupled with a certain finite, non-zero Δ(IN).

Now, if someone opens a window or door temporarily, it would be considered as an applied "disturbance", and the system will rush in to correct the resulting error. In that case, we can ask: what is the speed of correction? Or: what happens if we turn the knob of the thermostat a bit ("wiggling" the input of the system)? How quickly does the room temperature stabilize —called its "settling time". And so on.

In **Figure 1**, we present a basic control loop embodiment of this, as applied to a simple buck switcher for starters. With some differences, as explained below!

The IN node commonly used in general control loop theory is now the REFERENCE or REF node in power conversion. It creates the set point against which the output is compared. The closed loop gain of the system is now Δ(OUT) divided by Δ(REF). The OUT node in general control theory is V_{OUT} in power conversion. Despite some fairly common misinterpretation, the "IN" (of the control loop) is *not* the input to the power stage, or V_{IN}. It is the input to the control system, i.e. REF. Also, once the switcher is on, we don't really wiggle the thermostat/reference around! So the relevance of the oft-quoted closed-loop gain, as visualized and presented in textbook control loop theory, hardly has any significance to switchers. In switchers, the primary "inputs" we are interested in are essentially disturbances, injected at varying points within the closed loop system. Such as line and load variations. So we need to understand at a more general level, how disturbances are attenuated (hopefully not amplified), depending on their point of injection. It is not the same thing as wiggling a thermostat!

The "plant" or "process" in general control theory is now the entire block consisting of three cascaded stages: the PWM comparator, the switching stage, followed by the LC filter.

However, the "power stage" of the switcher by definition, traditionally includes only the latter two blocks. So it is the plant *less* the comparator. The comparator, though part of the plant, is considered part of the control section of the switcher since it contains no power components, just signal-level components.

The compensator in general control theory is typically an error amplifier, with all its feedback components present (usually a bunch of small-signal R's and C's). The "sensor" in general control theory (for example the thermocouple in the usual thermostat control loop theory example), is typically the voltage divider in power conversion. But more on its actual effect a little later.

The CONTROL terminal is the same in both representations. But in general control theory it may actually be called the "control effort", whereas in power conversion it is usually called the "control voltage" or "EA OUT" (error amplifier output).

Looking closely at **Figure** 1, we see that the entire process of control hinges on the concept of *negative feedback*. So if the output is going up, we try to quickly pull it down! That is why we see different signs around the summation block in the figure.

Note that in related literature, the summation block is often confusingly represented by a *multiplication* sign in the middle of a circle instead of a summation sign. That it is likely done only to make you give up control loop theory altogether, and leave it to the "experts"!

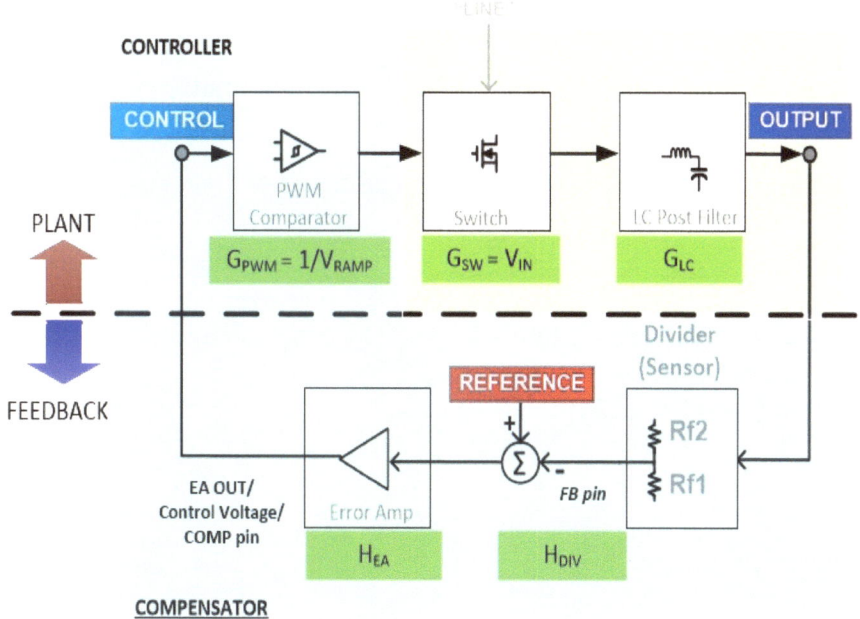

FIGURE 1: CONTROL LOOP OF A SWITCHER

Note that we will often use gain symbol "H" for parts of the feedback section, and "G" for the

plant. But in literature it is sometimes *the other way around,* with H being the plant and G the

compensator. Beware! Sometimes G is used for all the blocks, in the plant and the compensator.

Sometimes K is used, as in older Unitrode App Notes. Or "A" is used for the plant and β for the

compensator. And so on. Watch out for a lot of possible confusion as a result of all these

terminological variations.

The most important thing to keep in mind is that the representation of **Figure 1** assumes we have

a multiplicity of *cascaded* gain stages. Which implies the gain of each stage can be quantified as

a standalone, and the net gain is then the product of all the individual cascaded gain blocks. But

that may turn out to be a pipe dream. For example, the buck topology is the only one where we

can actually point to a separate "LC post filter" within the plant. In a boost or buck-boost, even

ignoring the relative locations of the L or C, the LC stage is really not "separable" from the rest—because the node between the L and C_{OUT} is connected to the switch/diode—unlike a buck. So we cannot separate the filter from the switch function. Well, at least not easily.

As pointed out in http://www.sciencedirect.com/science/book/9780123865335 , in the canonical model from Middlebrook (see http://ecee.colorado.edu/ecen4517/materials/Encyc.pdf), we can indeed separate the boost and buck-boost L and C into a separate LC stage, *provided we replace the inductor by an "equivalent inductor"* $\underline{L}$, equal to

$$\underline{L} = \frac{L}{\left(1-D\right)^2}$$

The C of this separated LC stage is still the original C_{OUT}. For a detailed discussion of how all the transfer functions of the non-buck topologies changes as a result of this, see

http://www.sciencedirect.com/science/book/9780123865335

The voltage divider is also not necessarily separable into a separate gain block. In **Figure 2**, we show how in a typical error amplifier stage, the lower resistor of the divider goes out of picture. It is just a DC biasing element, not connected with the AC response, which is what we are interested in. In other words, the divider doesn't enter the picture at all from a control loop a perspective. However, if we use a transconductance op-amp, the divider does enter the picture as separable gain block.

To experienced power engineers, some of the nuances mentioned above may seem to be subtle as flying quarter-bricks. It may genuinely surprise them to learn that these "details" are still often missed, or are at least routinely glossed over. But there are some other experts who are not at all surprised. They knew this was going to happen, and tried to warn budding power supply engineers since a very long time ago. But who was listening?

The unpolitically correct, or politically incorrect, gist of what some of them have said is: There are many self-proclaimed "experts" who *don't get it*. One of those memorable soothsayers was the flamboyant Lloyd Dixon of Unitrode (now Texas Instruments). That's exactly what he said while presenting his "Control Loop Cookbook" paper at the Unitrode Power Seminar in Germany. The year was 1996, and this author had the privilege to attend the presentations. Later that day, Bob Mammano, now considered the "father of the PWM IC industry", presented his topic: "Fueling the Megaprocessors - Empowering Dynamic Energy Management".

A short extract from Mr. Dixon's no-holds barred *written*, therefore more "PC" (politically correct), part of the presentation is reproduced below (see

http://encon.fke.utm.my/nikd/Dc_dc_converter/TI-SEM/slup113.pdf :

This is AC (change) analysis. Therefore V_{REF} is being ignored below as it is a biasing level only
By definition, transfer function ("H(s)") is output/input = V_{CONT}/V_O

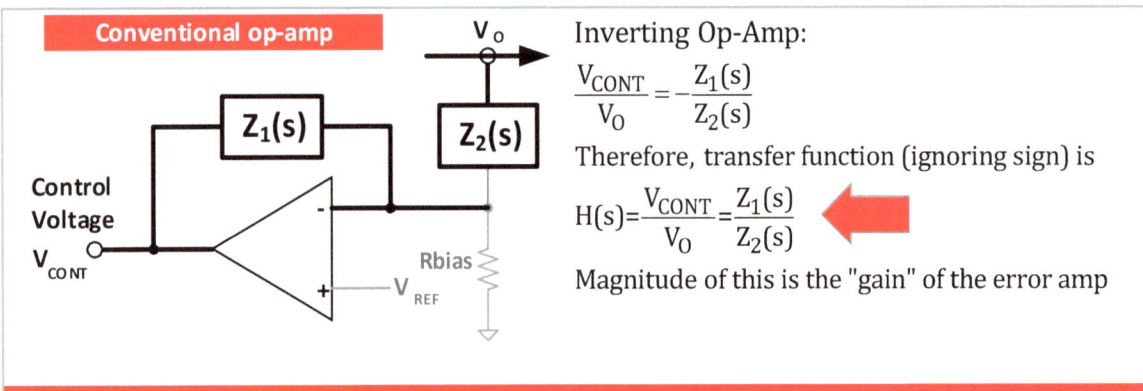

Conventional op-amp

V_O

Inverting Op-Amp:

$$\frac{V_{CONT}}{V_O} = -\frac{Z_1(s)}{Z_2(s)}$$

Therefore, transfer function (ignoring sign) is

$$H(s) = \frac{V_{CONT}}{V_O} = \frac{Z_1(s)}{Z_2(s)}$$

Magnitude of this is the "gain" of the error amp

Upper resistor Rf2 is in general Z_2 (s). The lower resistor Rf1 is "Rbias" here. However Rf1 is only a DC-biasing resistor --- it does not appear in the AC analysis, and is therefore not included in the transfer function above.

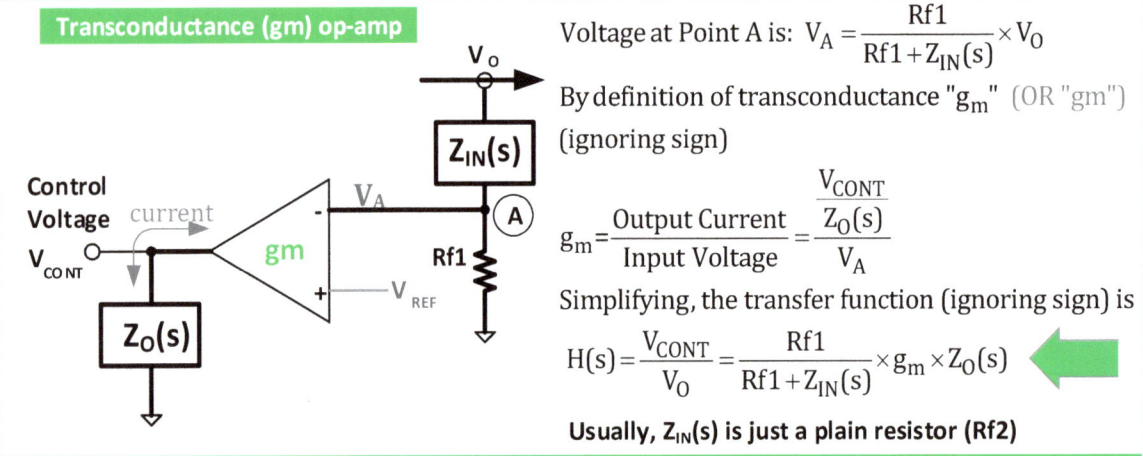

Transconductance (gm) op-amp

V_O

Voltage at Point A is: $V_A = \dfrac{Rf1}{Rf1 + Z_{IN}(s)} \times V_O$

By definition of transconductance "g_m" (OR "gm")
(ignoring sign)

$$g_m = \frac{\text{Output Current}}{\text{Input Voltage}} = \frac{\dfrac{V_{CONT}}{Z_O(s)}}{V_A}$$

Simplifying, the transfer function (ignoring sign) is

$$H(s) = \frac{V_{CONT}}{V_O} = \frac{Rf1}{Rf1 + Z_{IN}(s)} \times g_m \times Z_O(s)$$

Usually, $Z_{IN}(s)$ is just a plain resistor (Rf2)

Upper resistor Rf2 is in general Z_{IN} (s). The lower resistor Rf1 is not just a DC-biasing resistor --- it <u>does</u> affect the AC signal at the input, and through gm it affects the output current, and thereby the output voltage V_{CONT}. So Rf1 is included in the transfer function above.

FIGURE 2: IS THE VOLTAGE DIVIDER SEPARABLE AS A DISTINCT GAIN STAGE?

"Control Problems your Mother Never Told You About

A tremendous amount of effort has been put into the development of small-signal techniques and linear models of the various switching power supply topologies. Hundreds, if not thousands of papers have been written over the years. Your academic "mother", whoever "he" may be (note the PC sexual ambiguity), typically focuses on new topologies and/or linear modeling.

While not disparaging any of these efforts – far from it, these contributions have been immense and totally necessary – there has been a lack of balance and a tendency to try to force behavior that is uniquely related to switching phenomena into linear equivalent models (with sometimes uncertain results). Many of the major significant problems with switching power supplies do not show up in the frequency domain, or in the time domain using averaged models, unless these problems are anticipated in advance and provided for in the models. Simulation in the time domain using switched models, although slower, reveals these problems that would have been hidden." —*Dixon*

Perhaps somewhere along the way, a few relatively inexperienced engineers have gotten a bit carried away with their new-found prowess manipulating Laplace transforms and so on, and have therefore ended up downplaying, if not completely disregarding, some of these "subtle" aspects. Or perhaps their simulations failed to reveal what they perceived to be corner-case" problems— since they had used small-signal averaged or equivalent linear models for the switcher to start with, which then turned out to be a self-fulfilling prophecy: You can't see in the dark, if you didn't realize it may be dark and forget to bring along a flashlight.

The first thing we have to keep in mind is that control loop theory can be applied to a switcher *only when its power stage is considered reasonably optimal.* That should not become the stumbling block. Otherwise we would be just wasting all our valiant efforts in the frequency domain.

As an example, see **Figure 3.** Here we are showing a sample transient waveform from a vendor. See http://go.intersil.com/lp-stable-power-supply.html .

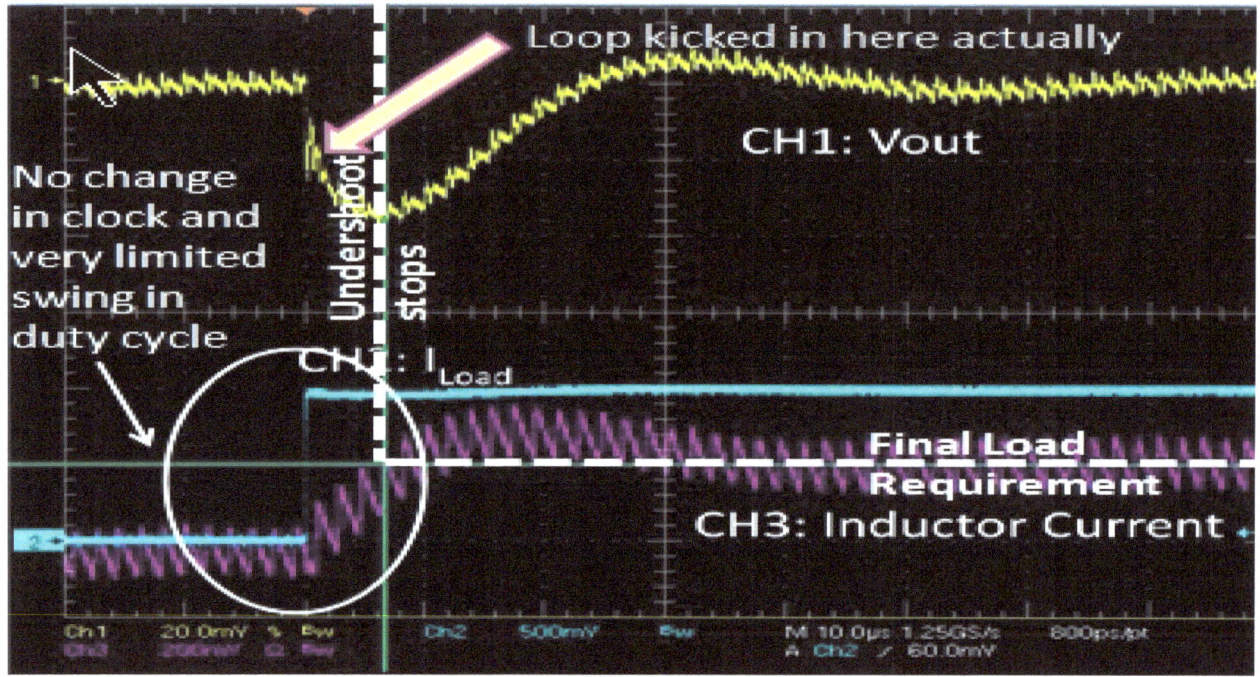

FIGURE 3: TYPICAL VENDOR PLOT SHOWING LOAD TRANSIENT AND INDUCTOR CURRENT (INTERSIL)

This vendor has helpfully provided the corresponding inductor current waveform too, which is quite unusual. It is quite revealing if you stare at it a bit.

The first interesting thing in **Figure 3** is: *the undershoot reverses direction exactly at the point where the inductor current reaches its final* intended *value.* So that seems to be the real gating item. *Not* the control loop. Of course, after that point, being a voltage-mode control system, the inductor current does overshoot a bit, commensurate with the observed output voltage overshoot. In a sense, the inductor current has a certain "momentum". But after a little ringing, the system stabilizes.

However, based merely on the fact that the *undershoot stops exactly at the point where the inductor current reaches its final* intended *value,* the overall response (undershoot) was *power-stage dominated,* not *control-loop dominated.* Which implies that anything more we could do with the control loop may have fallen flat on its face.

In fact the control loop appears to have reacted long before the maximum undershoot (minimum voltage) was recorded. See **Figure 3** again. That is actually typical of all well-designed control loops—they all "kick in" after about *three* or more switching cycles.

We see from the same figure that after the control loop kicks in, the control loop was certainly trying to command a correction, but oddly, nothing significant or dramatic happened—at least not immediately. Why? One reason for that could be some sort of an architectural limitation. Indeed, voltage-mode control (VMC) has some inherent deficiencies, based on the "momentum" of its inductor current. But despite that, VMC is nowadays considered a better choice, especially with *input feedforward* included (to be discussed shortly). That is in comparison to current-mode control, with its now-perceived inherent deficiencies such as subharmonic instability, noise sensitivity etc. That seems to be receding in the distance.

Note that hysteretic controllers seem quite promising in this regard, but they have a variable switching frequency, especially during transients. So, we may need to validate their system-level acceptability.

What did we mean about architectural limitations? Well, to optimize any voltage-mode controlled buck switcher, we need to, at a bare minimum, demand that it can get quite close to 100% duty cycle—to rapidly build up the inductor current as shown in Figure 4.

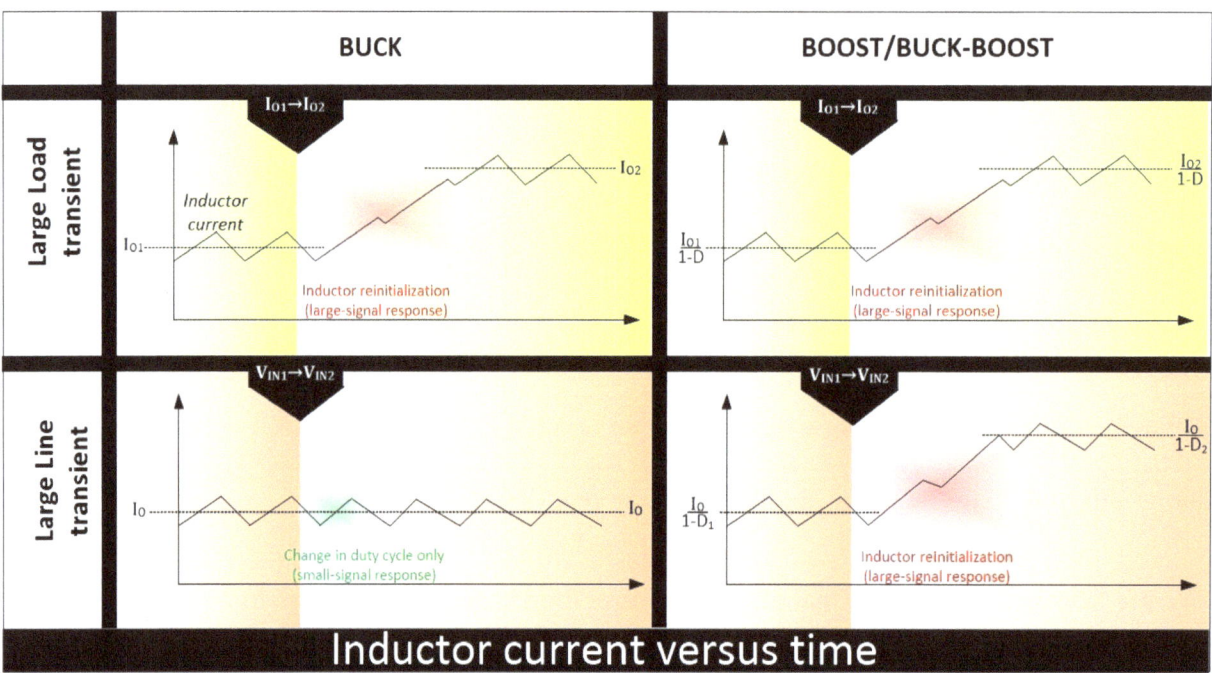

FIGURE 4: INDUCTOR "RE-INITIALIZATION" PROBLEM

Keep in mind that this ~100% duty cycle maximum limit is a very bad idea for a buck-boost or a boost, which actually *depend on the non-zero OFF-time to deliver energy to the output*! If we don't provide that, the output will continue to sag, whatever the current buildup in the inductor. And that is also the intuitive explanation of the oft-mentioned "right half plane (RHP) zero". The RHP zero doesn't exist for buck or buck-derived topologies such as the forward converter, but does enter the picture with the boost and with the buck-boost, including their derivative topologies such as the flyback.

Another thing we can do is change the clock to respond faster to a transient (pulse-on-demand), and if the varying clock frequency is acceptable, it should be fine. That's what happens in a hysteretic controller.

Unfortunately, from **Figure 3** we can see that there is no sign, either of on-demand pulses (off-cycle switch turn-ON), or of maximum duty cycle. Which could partially explain the slow curving change in output voltage, despite the control loop having kicked in.

Or as indicated, perhaps the inductance was simply too large.

However, it may also be that the power stage was well-designed and the duty cycle may indeed have been able to "max out" close to 100%, but it *simply wasn't asked* to, at least not quickly enough! Now, that would imply a poorly designed feedback stage, say with poor bandwidth! But we see enough signs that in this case the control loop did kick in after about three switching cycles. Which is about right! So, poor bandwidth doesn't seem to be the culprit here. A bunch of other things to check out instead.

The type of output response we would *like to see* is shown in the lower half of **Figure 5**. We don't need to see the inductor current to make some observations. Notice the sharp edge in the output voltage under/over shoot as compared to the waveform in the upper half of the figure. That seems to indicate a *control-loop dominated response*, not power-stage limited.

But also keep in mind that to declare victory, this particular sharp-edged output waveform must correspond to a *large*-signal event, such as zero to max load. We certainly cannot pass judgment on the power stage, whether it is "optimal" or "non-optimal", if we are only doing say, an 80% to 100% load test. Because then the small-signal/averaged models Mr. Dixon warned us against, do apply. In effect, we are then no longer dealing with real-world switchers. Just textbooks.

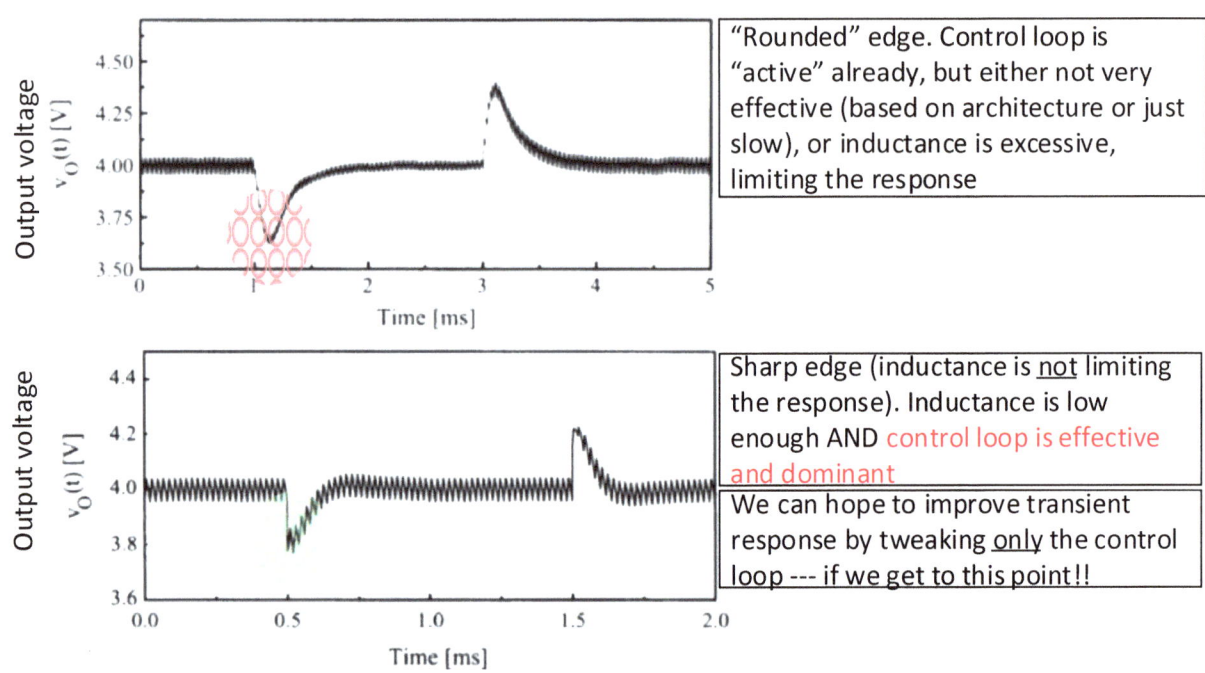

FIGURE 5: THE EDGE OF THE OUTPUT RESPONSE REVEALS A LOT

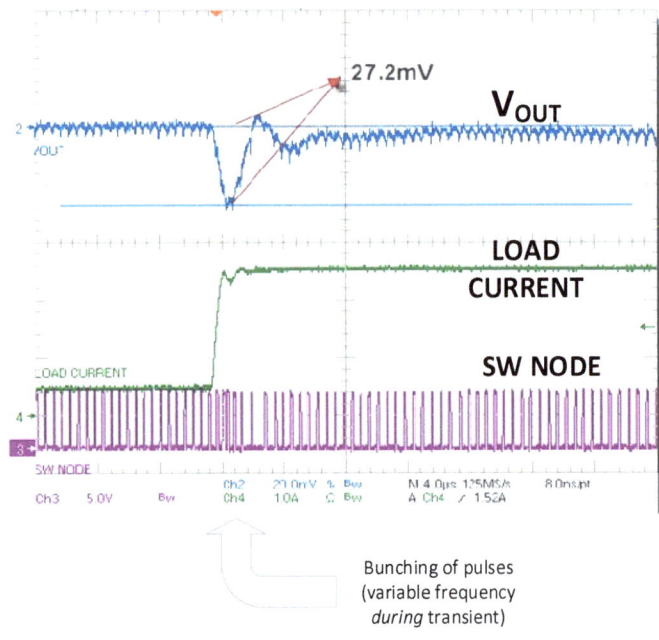

FIGURE 6: HYSTERETIC CONTROLLER RESPONSE

26

In **Figure 6** we show the exact response of a typical hysteretic controller. Notice the on-demand pulses here, and also the sharp edge as the undershoot starts to head upwards.

So, one thing seems quite certain. To get the control loop to make its presence felt during large signal events, we need to optimize the power stage first. One of those steps is to lower the inductance. However, placing the current ripple ratio "*r*" to be close to 0.4 as advocated by this author, is definitely in the ball-park, and there is likely no point reducing the inductance any further. See www.ti.com/cn/lit/pdf/zhca135 and www.ti.com/lit/an/snva038b/snva038b.pdf, originally written by this author in 2001 at National Semiconductor.

What about the output capacitor?

We now hark back to the solved example in Chapter 19 of *Switching Power Supplies A-Z, Second Edition,* to reveal the importance of selecting this vital component too. The results are presented in **Figure 7**. Basically, there are three main criteria for capacitor selection. One is the ripple, measured under steady-state max load. It leads to a minimum capacitance requirement of 5.2 μF in this particular example. Another is based on the overshoot which will occur if we just suddenly disconnect the load. This leads to a minimum capacitance requirement of 22 μF. And third, we have a certain *control loop assumption*, which says the output cap must be capable of providing *all* the energy for at least *three* cycles, in case of a large load step. Because during these three cycles, in effect the inductor is not capable of providing most of the energy requirement, as its current slews up in accordance with **Figure 4** (discussed in more detail later). This leads to a minimum capacitance requirement of 30 μF in our example. We picked 33 μF as a potential final value. We should however consider tempcos (temperature coefficients) again, and voltage coefficients too.

Note that if in the second calculation, instead of a 2.2 µH inductor, we had used a 4.7 µH inductor, it would have required a minimum capacitance of almost 50 µF, which would have overshadowed the control loop dictated minimum capacitance of 30 µF. *So, we have to be very careful not to select a larger inductance than recommended.* But $r = 0.4$ should work fine usually.

On the other hand, if the control loop design is sluggish, requiring not 3, but say 6 switching cycles to start acting, we would need to correspondingly increase the minimum capacitance from 30 to 60 µF. That adds to the cost too.

So, hand-in-hand with power stage optimization, we need to optimize the control loop too for best results.

And finally, besides juggling a few things around before deciding the optimal power component selection, let's not forget the basic architecture of the controller either.

Minimum Capacitance and Maximum ESR based on maximum output ripple

Ignoring ESR and ESL, purely based on capacitance, the maximum allowed output ripple determines a minimum output capacitance.

$$C_O \geq \frac{r \times I_O}{8 \times f \times V_{RIPPLE\_MAX}}$$

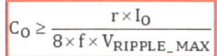

Including ESR, but assuming C is large and ESL is negligible. The maximum allowed voltage ripple determines a maximum ESR

$$V_{RIPPLE} = ESR \times I_O \times r$$

$$ESR \leq \frac{V_{RIPPLE\_MAX}}{I_O \times r}$$

Minimum Capacitance based on maximum overshoot

There is another criterion. In case of a sudden release of load demand, say from max load Io to zero, the inductor energy will all get dumped into the output cap. If we do not want too much of an overshoot (to a new value Vx):

$$\frac{1}{2} \times C\left(V_X^2 - V_O^2\right) = \frac{1}{2} \times L\left(I_O^2\right) \quad \Rightarrow \quad C \geq \frac{L\left(I_O^2\right)}{(V_X + V_O) \times (V_X - V_O)} \approx \frac{L\left(I_O^2\right)}{(2V_O) \times (\Delta V_{overshoot})}$$

$$C_O \geq \frac{L\left(I_O^2\right)}{(2V_O) \times (\Delta V_{overshoot})}$$

(where we have used the approximation $V_X + V_O \approx 2 \times V_O$. Also, $V_X - V_O = \Delta V$)

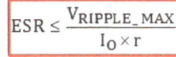

Minimum Capacitance based on maximum droop

Typically, with a well-designed loop, it takes about **three switching cycles for the loop to react** and start correcting the output to meet a sudden load demand. During that time we do not want the output cap to fall more than a certain value Vdroop. Thus, using I = C dV/dt, we get

$$I = C\frac{\Delta V}{\Delta t} \quad \Rightarrow \quad C \geq \frac{I \times \Delta t}{\Delta V} = \frac{I \times 3T}{\Delta V_{droop}} = \frac{I \times 3}{\Delta V_{droop} \times f}$$

Here the droop is actually related to the <u>extra</u> load demand, since the normal load requirement is being met every cycle without any droop. So the current here is actually the load increase.

$$C_O \geq \frac{3 \times \Delta I_O}{\Delta V_{droop} \times f}$$

$$C_{O\_MIN\_1} = \frac{r_{VINMAX} \times I_O}{8 \times f \times V_{O\_RIPPLE\_MAX}} = \frac{0.4147 \times 5}{8 \times 10^6 \times 0.05} = 5.1834 \times 10^{-6} \text{ F} \qquad \text{5.2 μF} \quad \text{#1}$$

$$C_{O\_MIN\_3} = \frac{L \times I_O^2}{2 \times V_O \times \Delta V_{OVERSHOOT}} = \frac{2.2 \times 10^{-6} \times 5^2}{2 \times 5 \times 0.25} = 2.2 \times 10^{-5} \qquad \text{22 μF} \quad \text{#2}$$

$$\qquad\qquad\qquad\qquad\qquad\qquad\qquad\qquad\qquad\qquad\qquad\qquad\qquad\qquad\qquad \text{30 μF} \quad \text{#3}$$

$$C_{O\_MIN\_2} = \frac{3 \times (I_O/2)}{\Delta V_{DROOP} \times f} = \frac{3 \times (5/2)}{0.25 \times 10^6} = 3 \times 10^{-5}$$

We picked 33μF standard value

(but need to watch out for tolerance, temperature, voltage coefficients etc...may need to oversize as much as by a factor of 2)

Note: If Inductance was 3.3uH, not 2.2uH we would get 33 uF not 22uF for condition 2 above. It will therefore start dominating the selection! **Ensure inductance is NOT excessive**

FIGURE 7: WORKED EXAMPLE FOR OUTPUT CAPACITOR SELECTION CITERIA

Referring to Lloyd Dixon's presentation again, he says: "*The open-loop gain, T, is defined as the total gain around the entire feedback loop (whether the loop is actually open, for purpose of measurement, or closed, in normal operation).*" So in our terminology, T = GH, where G and H themselves may be the product of cascaded stages as we can see from **Figure 1**.

Similarly, Dixon says: "*Closed-loop gain defines the output vs. control input relationship, with the loop closed*".

Actually Mr. Dixon is calling the reference as the "control" node here. Which is a bit misleading. Besides that, the reference in his case is placed between the voltage divider and the error amplifier. Which actually need not be the case always, even assuming the divider can be taken out as a separated gain stage, which in reality may not be so either, as mentioned previously.

Which is why it is very important to truly understand how a disturbance gets attenuated on account of the closed loop control system, compared to the case of no feedback (open control loop). And this also depends on the *point of injection of the disturbance*, in this case the "wiggle" in the reference (which as mentioned is actually of no real significance in a switcher either!).

Just to resolve the widespread confusion, let's see the form of the open-loop gain function and get comfortable with our understanding.

Refer to **Figure 8**, where we compare what happens in two cases, depending on the location of the reference wiggle, whose effect on the output is what closed loop gain is all about.

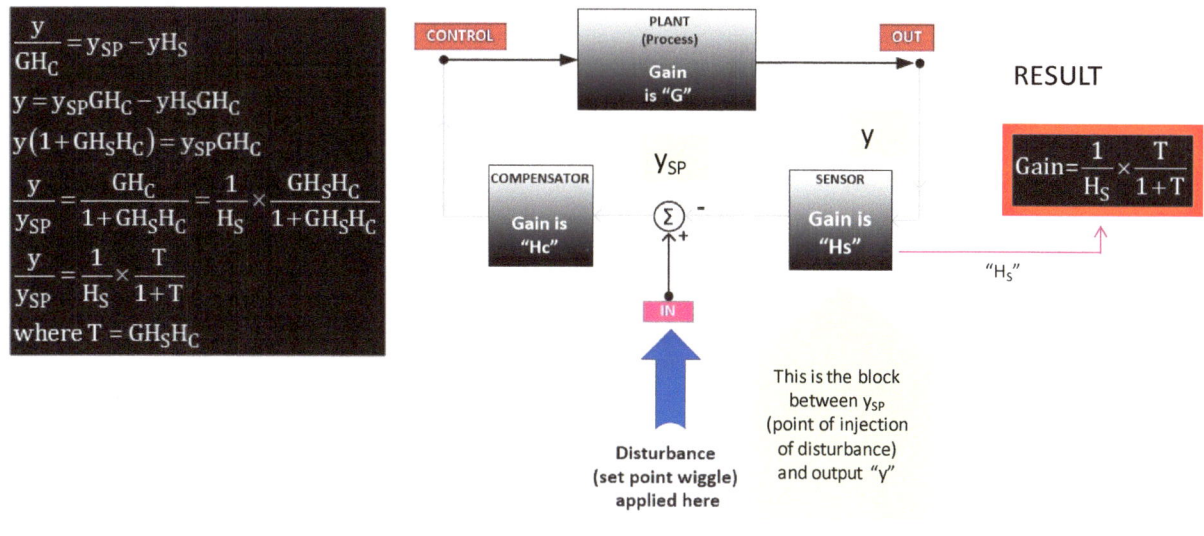

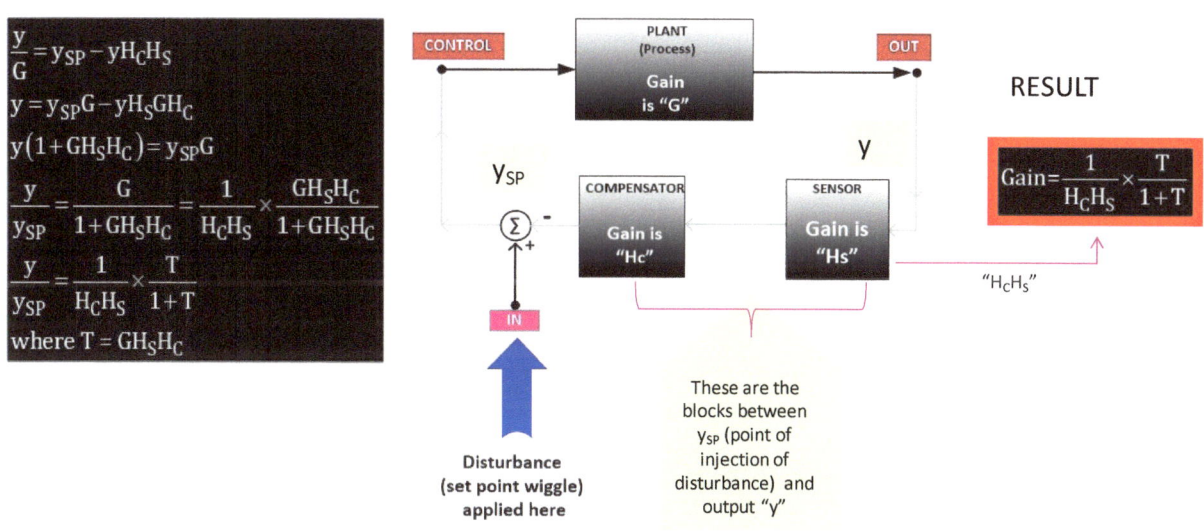

FIGURE 8: POINT OF INJECTION OF DISTURBANCE AND CORRESPONDING CLOSED LOOP GAIN

In the first (top) case we first realize that any change in the output, "y", propagates clockwise in the close-loop system. So starting from the output, we retrace its path *backwards* through the plant G and the compensator H_C (anticlockwise), and the signal at the output of the summation block must be $y/(G \times H_C)$. Now, going clockwise from the output rail instead, y becomes $y \times H_S$ after passing through the sensor. After the summation block it is therefore $y_{SP} - yH_S$. But this must equal $y/(G \times H_C)$. Equating the two, we get the expression evaluated within the figure.

31

Similarly, we can go clockwise and also anticlockwise with the reference re-positioned as shown in the lower half of **Figure 8**, and we get the expression evaluated within the figure.

We see that in both cases of the figure, we end up with a form

$$\text{Open Loop Gain} = \frac{y}{y_{sp}} = \frac{1}{H_x} \times \frac{T}{1+T}$$

where H_x is the net gain of all the gain blocks between y and y_{sp} in the forward (clockwise) direction.

Note that T is simply the net gain of all the blocks involved, be they considered part of the plant (G) or of the feedback network (H). And that leads us to a very general derivation of the closed loop gain of several cascaded stages, where we can just use one symbol for all of them, say G as in **Figure 9**.

What this figure is saying is that:

$$\text{Gain with feedback} = \text{Gain with no feedback} \times \frac{1}{1+T}$$

In terms of conventional labels, *"Gain with feedback" is the closed loop gain.* In fact "Gain with no feedback" should actually be considered as the open-loop gain, since that expresses the actual change in the output for a change in REF, with no feedback present. If that is so, *then T should more correctly be simply called the "loop gain"* since it is the cascaded product of all the gain stages of a closed loop system.

$1/(1+T)$ is the "correction factor" which tells us that the effect of closing the loop reduces (hopefully) the effect of the disturbance on the output, by the factor $1/(1+T)$.

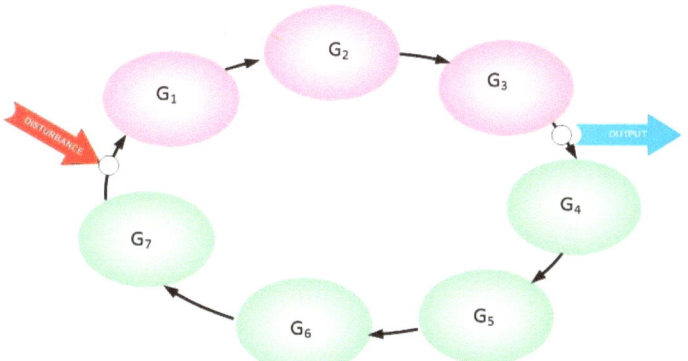

In general, IN can be any input disturbance (line, reference etc.) injected from outside into the closed loop of the plant and feedback.

Its effect on the output OUT is:

$$\left.\frac{OUT}{IN}\right|_{no-feedback} = G_1 G_2 G_3 \qquad \text{\textbf{Gain with no feedback}}$$

$$\left.\frac{OUT}{IN}\right|_{with-feedback} = \left(\frac{1}{G_X}\right) \times \left(\frac{T}{1+T}\right) = \left(\frac{1}{G_4 G_5 G_6 G_7}\right) \times \left(\frac{G_1 G_2 G_3 G_4 G_5 G_6 G_7}{1+T}\right) = G_1 G_2 G_3 \boxed{\left(\frac{1}{1+T}\right)}$$

Where T= "loop gain" (i.e. T = G$_1$ x G$_2$ x G$_3$ x...x G$_N$)   **Correction factor**

FIGURE 9: GENERAL FORM OF CLOSED LOOP GAIN FOR ARBITRARY POINT OF INJECTION OF DISTURBANCE

We can do a more detailed analysis of the correction factor:

a) For DC, or rather near-DC, where there is no or little associated phase shift, T is just a real number (no imaginary component). And if it is very large, then $1/(1+T) \approx 1/T$. *So it reduces the effect of the disturbance on the output by the factor 1/T, compared to the case where there was no feedback present.* At least for low frequencies, this is easy to visualize.

For example, if in a buck, the input is 10V and output is 1V, the output is $V_{OUT} = V_{IN} \times D$. So if the input doubles to 20V, so will the output (with no closed loop in place). In terms of dB, the input doubled ($20 \times \log 2 = 20 \times 0.3 = 6$ dB), and the output also went up by 6 dB. However in AC analysis, gain is the ratio of the *change* in output to a *change* in input. The correct way to proceed is shown in **Figure 10**.

$$\text{Input\_Output\_Gain}_{\text{with\_no\_feedback}} = 0.1$$

So if input jumps by 10V (to 20V), the output will jump by $10V \times 0.1 = 1V$ (to 2V)!

$$\text{Input\_Output\_Gain}_{\text{with\_no\_feedback\_dB}} = 20 \times \log(0.1) = -20\text{dB}$$

$$\text{Correction Factor} = \frac{1}{1+T} = \frac{1}{201} = 5 \times 10^{-3}$$

$$\text{Correction Factor}_{\text{dB}} = 20 \times \log(5 \times 10^{-3}) = -46\text{dB}$$

$$\text{Input\_Output\_Gain}_{\text{with\_feedback\_dB}} = \text{Input\_Output\_Gain}_{\text{with\_no\_feedback\_dB}} + \text{Correction Factor}_{\text{dB}}$$

$$\text{Input\_Output\_Gain}_{\text{with\_feedback\_dB}} = -46 - 20 = -66\text{dB}$$

$$\text{Input\_Output\_Gain}_{\text{with\_feedback}} = 10^{\text{dB}/20} = 10^{-66/20} = 5 \times 10^{-4} = 0.5\,\text{m}$$

So now, if input jumps by 10V, the output will only jump by $10V \times 0.5\,\text{m} = 5\,\text{mV}$

Higher T will reduce this error further.

FIGURE 10: SAMPLE CALCULATION OF DC SETTLING ERROR

Note that when we express gain in terms of decibels, multiplication (say of cascaded gain stages) is reduced to a summation (in decibels). *Also, here we are referring to the "input" as literally the input rail, not the reference.* We show how the expected 1V shift in the output on account of the input doubling, with no closed loop correction, is reduced to just 5 millivolts on account of the high DC gain of the system.

b) At high frequencies, if T equals -1, the denominator will "explode". Which means that we will in effect have sustained oscillations, because there are always limiting parasitics present, which will not allow the output to really rise to "infinity".

Still, we need to understand how T can equal -1. Very simply that means, in terms of magnitudes and phase expressed in polar notation, i.e. in the form of $r \angle \theta$, we can write $T = 1\angle\text{-}180°$. That is a magnitude of 1 and opposite phase.

34

So we have full-blown instability if the loop gain T equals 1, and the corresponding phase is -180°. Why is that a problem? Because combined with the intrinsic -180° associated with negative feedback (the signs around the summation block where the reference is introduced in **Figure 1**, or the fact that the output is fed to the *inverting* pin of the op-amp in **Figure 2**), we get a total phase lag of -360°. This means: "in-phase". The disturbance is reinforcing itself.

We thus arrive at the criterion for instability of a closed-loop system: It all depends on T, which as we are now realizing is better referred to as the "loop gain", not really "open-loop gain", which is rather misleading. So at the particular frequency at which T = -1, the disturbance has in effect gone around the closed loop and returned to the point of injection with exactly the same magnitude and phase it started off with. So, it is going to reinforce *and* sustain itself!

The frequency at which $||T|| = 1$ (same as 0dB axis) is called the crossover frequency. If at the crossover frequency, the phase of T is exactly -180°, then the system will be unstable!

Note that in the 1996 Unitrode presentation, Mr. Dixon went to great pains to explain why there is a possibility of oscillation *only at the crossover frequency*. Why is it that the signal can actually return in phase with a gain far greater than 1, and the system still be considered stable? Mr. Dixon confessed to spending "sleepless nights" thinking about this, and ultimately explained it as a vector formation that just can't "close"…and thus can't exist, at least not for long.

But it does throw up the possibility of "conditional stability" which people in the industry seem to have widely differing views on. Because one problem is, if during a sudden large load transient, the inductor needs time to build up current, or the error amplifier "rails", then in effect the gain collapses, and it could at some point reach the self-sustaining condition expressed as T $= 1\angle -180°$.

This author will at a later stage describe how in fact this "conditional stability", which incidentally, even Ray Ridley underplays completely (see http://www.ridleyengineering.com/loop-stability-requirements.html?showall=&start=2) can actually be a major contributor to the "ringing" we see on the output during load transients, *and thus should be carefully weeded out as far as possible.*

Lloyd Dixon also cautioned quite a bit against conditional stability, but more in relation to the gain collapsing for various reasons, and the resulting propensity for sustained oscillations. He didn't relate it to any improved transient response, which we will do later in this book.

In an effort to create a safety margin from instability, the terms "phase margin" and "gain margin" were coined. See **Figure 11**. This is called the Bode plot and it tells us everything about "T", the loop gain (previously called open loop gain). That is all we need to know for ensuring stability (Nyquist's criterion, as Mr. Dixon pointed out, is useful, but truly necessary only in cases where there are multiple f_{CROSS}).

The limitations of Bode plots are: they really tell us nothing about what *will* happen in the *time domain*—in terms of amplitude or frequency of overshoots or undershoots as a result of any line or load transients, or even reference-voltage wiggles. That is why the "connection" to any optimum phase margin remains nebulous.

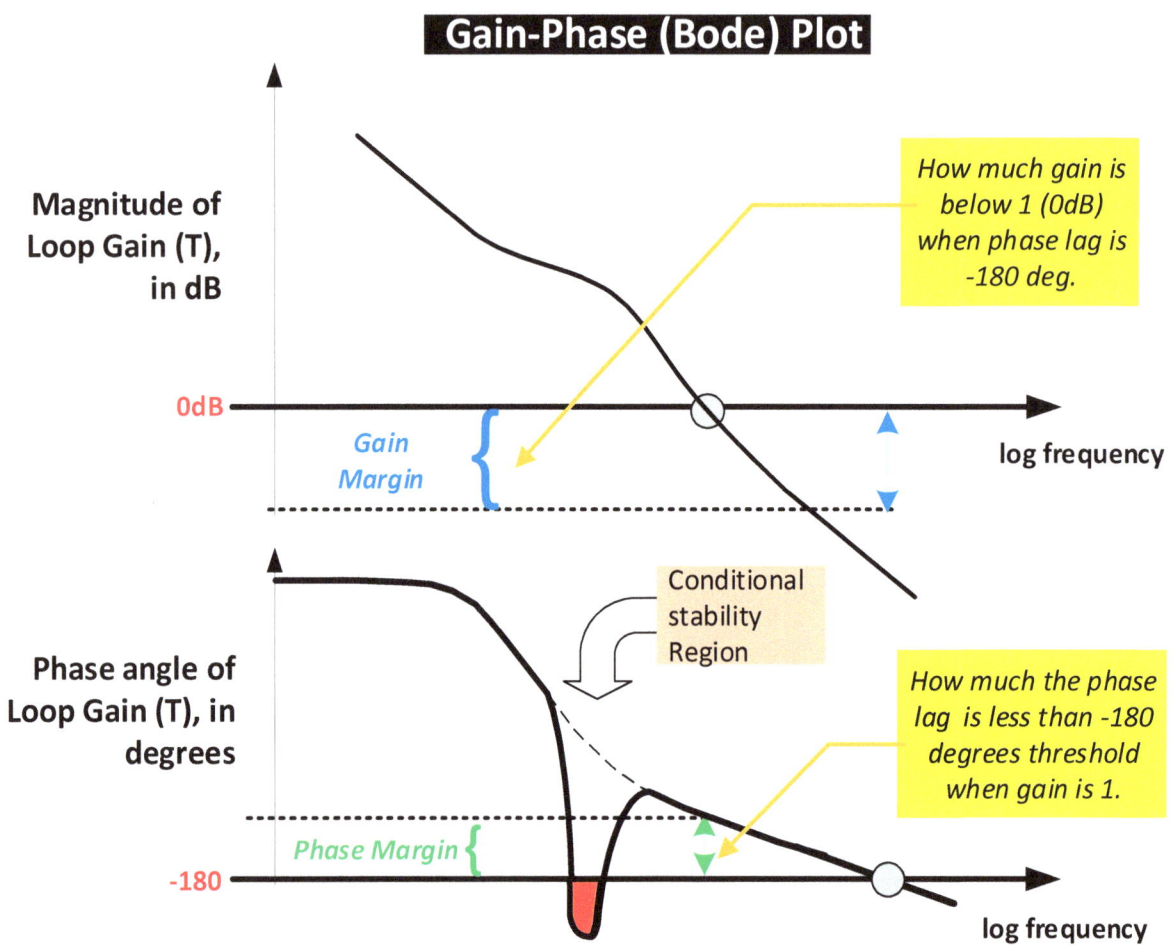

With an inherent 180 degrees phase lag on account of negative feedback, if the the total phase lag reached 360 degrees, with a gain of unity, a sustained oscillation becomes possible

FIGURE 11: GAIN AND PHASE MARGINS, ALONG WITH CONDITIONAL STABILITY

In **Figure 11,** we also see what conditional stability can look like, but it is not clear from the Bode plot what all are the unintended consequences of this—whether it should be subdued in some way, or if it truly represents a "rugged" system as Ridley opines here:

http://www.ridleyengineering.com/loop-stability-requirements.html?showall=&start=2.

Gain and phase margin are inter-related based on the fact that we typically aim for the loop gain T to *drop at the rate of "-1"*. That is a slope of -20 dB/decade. Note that gain expressed in decibels (dB) is $Gain_{dB} = 20 \log \|Gain\|$. So this means the gain is falling at the rate of 10× (i.e. 20 dB) for a 10× (i.e. decade) shift in frequency. This simply means we have set the loop gain to be *inversely proportional to the frequency* past some break-point (in this case the break-point is close to 0 Hz). That is the most common and easily-handled profile for T, because it corresponds to a "first order filter" (involving only one reactive component combined with a resistance). Such filters can produce only 90° of phase shift, so that leaves us with a comfortable (stable) phase margin of around 180-90 = 90°. Though with some reactive parasitics added to it, we may end up with a lower phase margin. Or we may deliberately try to lower the phase margin, say by placing a pole at a frequency close to the crossover frequency, and so on. But a second-order filter profile for T would have produced a 180° of phase shift right off the bat, rendering it unusable, because the phase margin would then be zero! So a "-1' slope is what we aim for.

Table 1 Phase Margins and Gain Margins

Phase Margin (Degree)	Gain Margin (dB)	
20	3	Great ringing, absolutely bad value
30	5	Slight ringing, slightly bad value
45	7	Borderline damping, best response time
60	10	Generally appropriate value
72	12	Reference value, no peak in the closed loop response

Courtesy: FRA App Note, NH Corporation Japan

FIGURE 12: RELATIONSHIP BETWEEN GAIN AND PHASE MARGIN (TYPICAL)

In **Figure 12**, we show a typical table of the relationship between phase and gain margin, based on the "-1" slope profile we desire. However, some reactive parasitics could cause the phase to veer upwards as shown in **Figure 13**, and so there may be no way, practically speaking, to define a gain margin. In that case, ensuring phase margin should usually suffice.

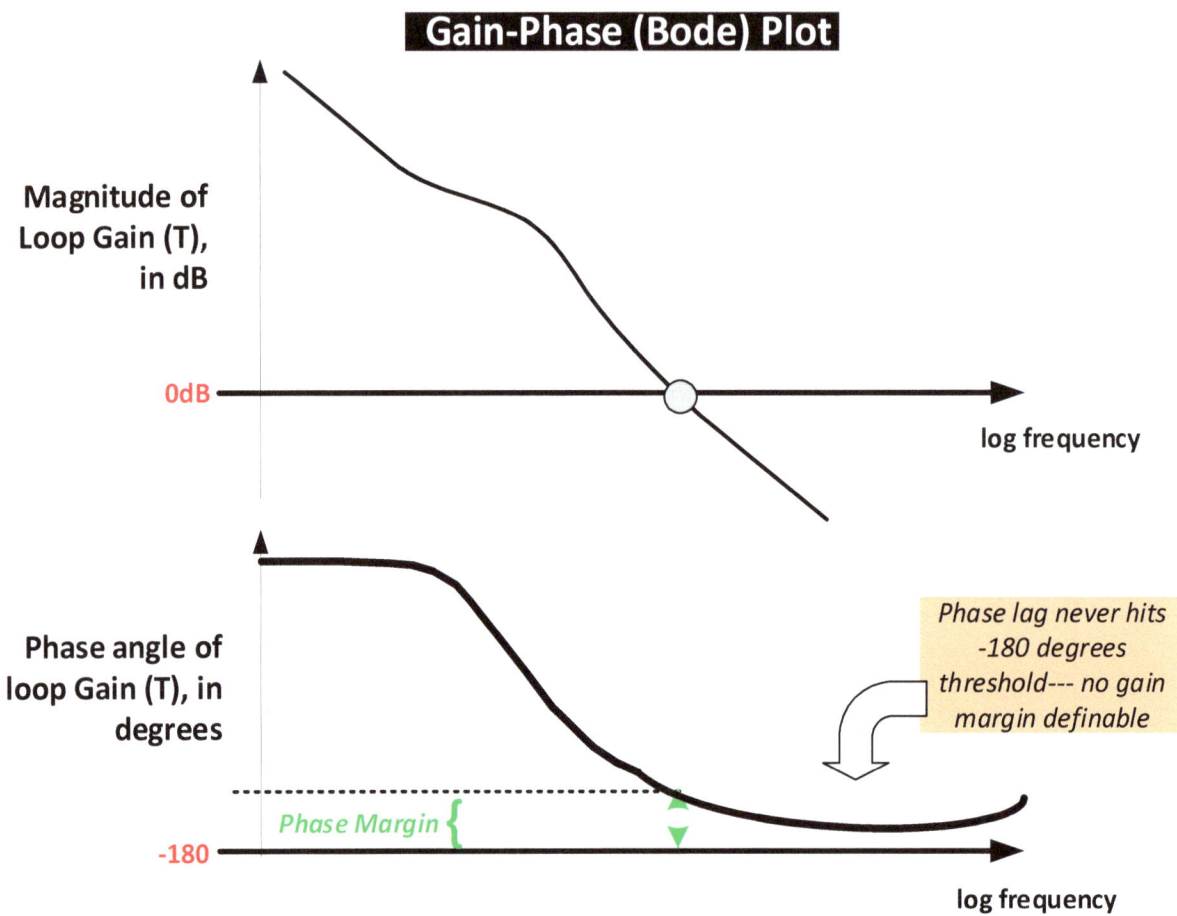

FIGURE 13: GAIN MARGIN MAY NOT BE DEFINABLE

We typically pick a high DC gain for T and then roll it off at high frequencies to avoid phase-shift based reinforcement of disturbances. That defines the AC response. But it is reassuring to confirm how a high DC gain helps in bringing the DC rail close to its intended set point value (reference).

In **Figure 14**, we show how this works, putting some numbers to the test to feel comfortable. It reveals that *a certain settling error is inevitable*, since DC gain is practically never infinite.

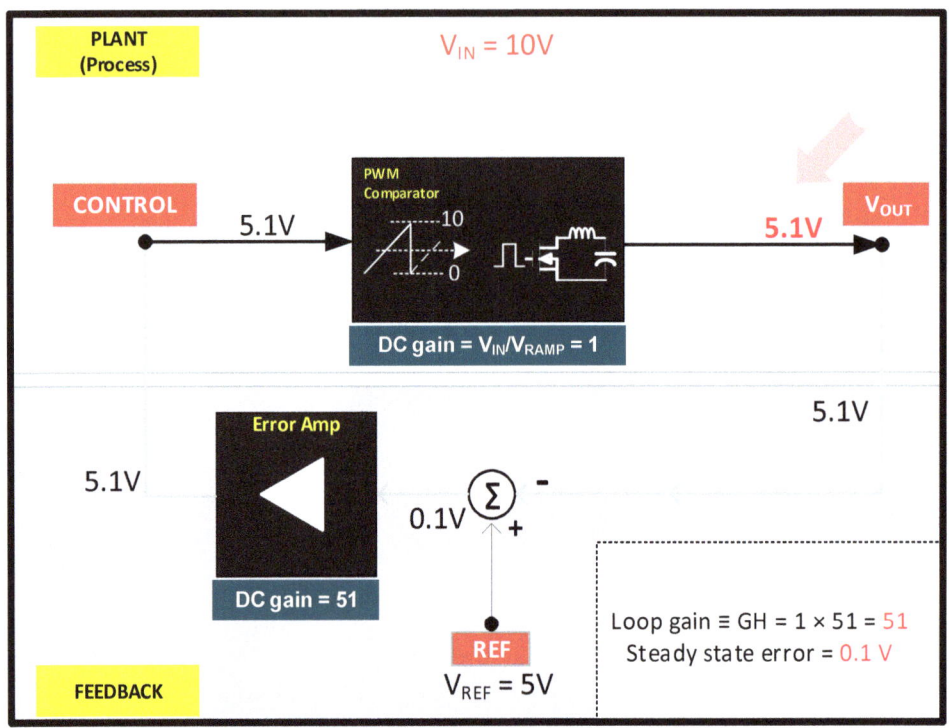

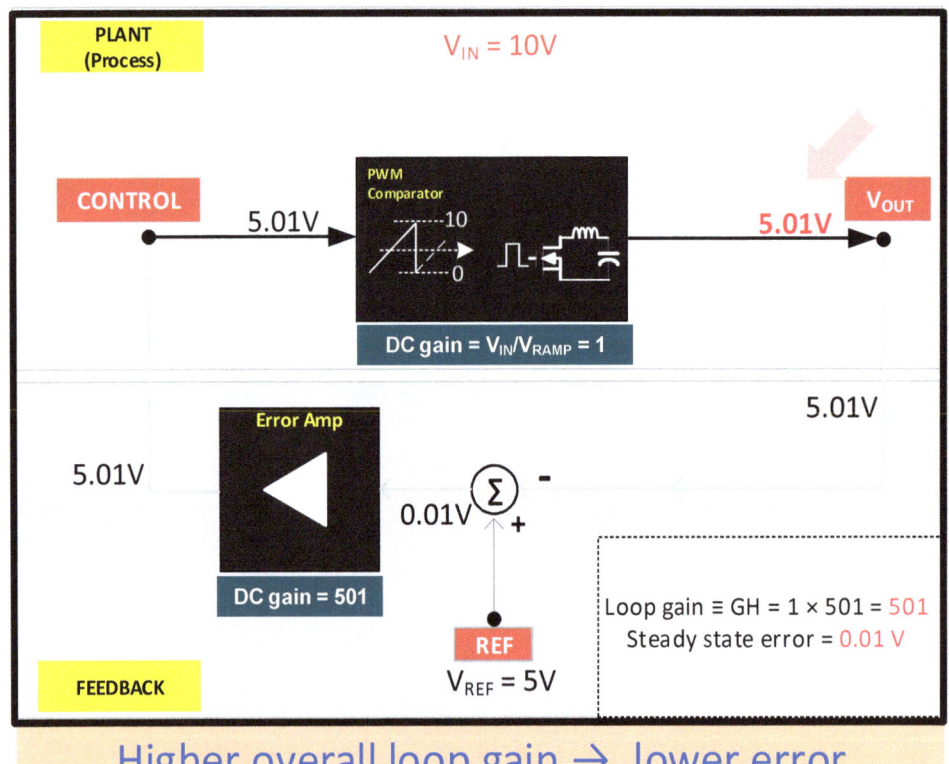

FIGURE 14: DC GAIN AMD SETTLING ERROR EXAMPLE

This leads to the AC and DC analysis shown in **Figure 15**, which serves to emphasize what we are really trying to eventually do. It is good to keep our key goal in mind at all times, lest we lose

our way.

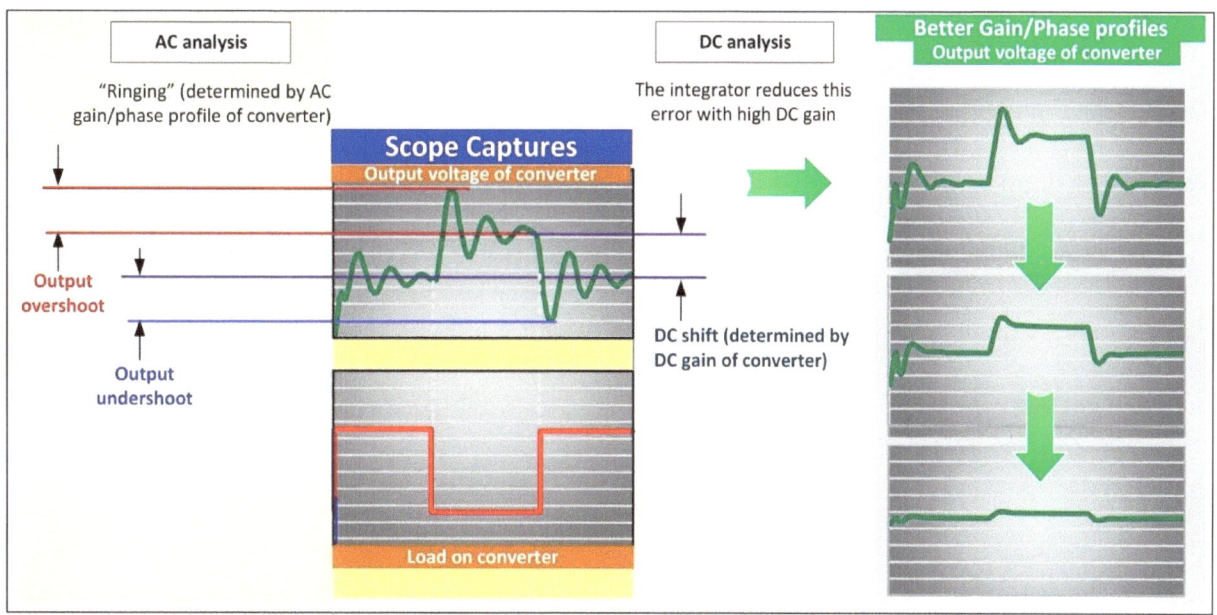

FIGURE 15: AC AND DC ANALYSIS OF A TRANSIENT WAVEFORM AND ULTIMATE GOAL

VOLTAGE POSITIONING

There is a technique to improve AC response, by trading off some DC gain in the process. Though that leads to a deterioration in the settling accuracy, as we see from **Figure 16**, it helps restrict the transient response to within a certain acceptable window.

This technique essentially allows the output rail to collapse a little bit as the load is increased. This can be done by introducing a small resistance after the point of regulation and between the load. That would be called passive voltage positioning. Since this is bound to be slightly dissipative, the modern technique actually varies the set point as a function of load. It is called *active voltage positioning.*

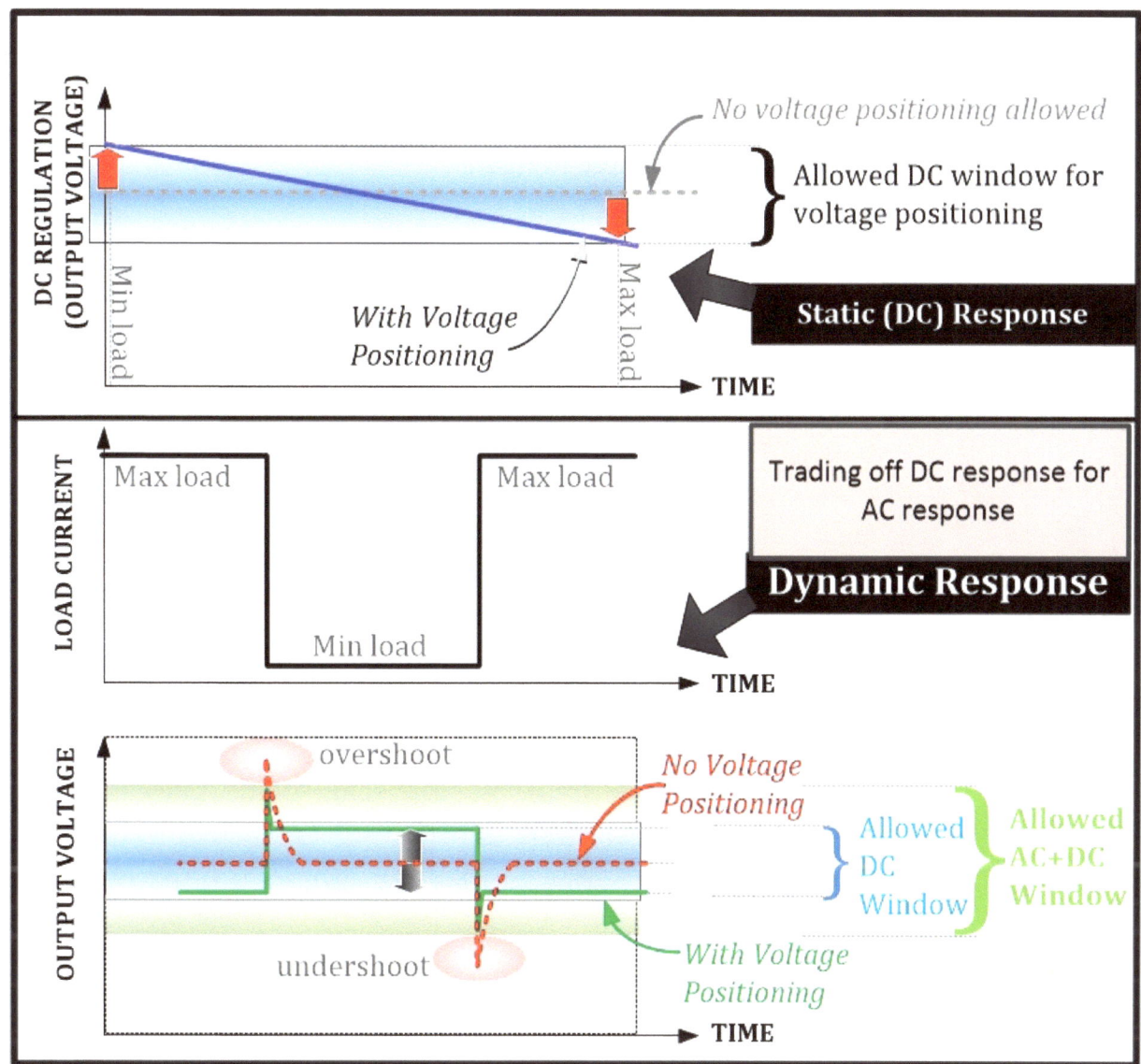

FIGURE 16: VOLTAGE POSITIONING

Either way, this positions the output voltage at the lower end of the acceptable window, so now

if we suddenly remove the load, the output rail will tend to fly up as expected. *But since it was*

positioned further down to start with, it now has a larger available overshoot (excursion) before

it exceeds the upper threshold of the allowed window.

Some have expressed the view that this allows the output capacitance to be reduced somewhat.

See www.linear.com/docs/5600. Indeed, but only if the control loop is the dominant criterion for

selection of the output capacitance, as described in **Figure 7**.

Returning to **Figure 10**, we featured a simple numerical example using a regular buck converter to show how a high DC gain helps reduce the effect of input variations on the output. Now we want to extend the same argument to an AC-DC power supply and show how the crossover frequency attenuates the low-frequency input voltage ripple from appearing on the output.

Note that here, *"input" once again refers literally to the input rail, not the reference.* And admittedly, it is better to call that the "line" instead, to avoid confusion.

Let us take the case of an AC-DC power supply with a certain input ripple at 100Hz (full-wave rectified input of 50 Hz). Assuming it is a forward converter with buck-like characteristics, and its duty cycle is 30%, the input-to-output transfer function will provide a dc attenuation of $|20 \times \log(D)| = 10.5$ dB, because D is the factor that connects the input rail to the output rail, as in **Figure 10**.

But this may receive a further attenuation due to the turns ratio, which may be $N_{PRI}:N_{SEC}$ equal to 20:1. That gives us $20 \times \log(20) = 26$ dB. So we have a net attenuation, without feedback, of 10.5 + 26 = 36.5 dB. In terms of actual factors, this is equal to an attenuation of

$$Gain\_attenuation = 10^{dB/20} = 10^{36.5/20} = 66.8$$

This means that if we have an input ripple of 10V, the output would have seen a corresponding ripple component of 10/66.8 = 150 mV.

But now let's introduce closed-loop correction. Suppose the entire loop gain (T) is such that it falls at roughly -1 slope, and crosses over at 50 kHz. We ask: what is the loop gain at 100 Hz? —the frequency of our interest here. Since a -1 slope simply indicates inverse proportionality,

$$\frac{\text{Loop\_gain}_{100\text{Hz}}}{\text{Loop\_gain}_{\text{fcross}}} = \frac{f_{\text{CROSS}}}{100\text{Hz}} \text{ (since -1 slope implies inverse proportionality)}$$

$$\text{Loop\_gain}_{100\text{Hz}} = \frac{50000}{100} = 500 \text{ (since Loop\_gain}_{\text{fcross}} = 1 \text{ by definition)}$$

Expressed in dB, this is

$$20 \times \log\left(\text{Loop\_gain}_{100\text{Hz}}\right) = 20 \times \log(500) = 54 \text{ dB}$$

So since the correction factor is $1/(1+T) \approx 1/T$, this is equivalent to an additional attenuation of 54 dB. So now the net attenuation is 54 + 36.5 = 90.5 dB. In terms of factors

$$\text{Gain\_attenuation} = 10^{\text{dB}/20} = 10^{90.5/20} = 33.5\text{k}$$

This means that if we have an input ripple of 10V, the output will see a corresponding ripple component of 10V/33.5k = 0.3mV.

This is a major improvement over the 150mV without feedback.

Of course, to get the actual output ripple, we have to add the contribution from the output filter stage etc. This is just the additional low frequency modulation that will be superimposed on that high-frequency ripple.

The PWM comparator is a key gain block of the closed loop system, as shown in **Figure 1**. Its gain (transfer function) has an input which is the control voltage, and an output which is the duty cycle.

The PWM comparator basically superimposes the control voltage against a ramp, and picks a duty cycle based on the intersection, as shown in **Figure 17**. Since the control voltage is the "in", and D is the "out" for this gain block, we can see from the figure that the gain is simply $1/V_{RAMP}$. Smaller the ramp, higher the gain. Also, this gain is not a frequency-dependent block. It applies to all frequencies extending up to the switching frequency and beyond. There is no "associated" phase shift either. This is just "DC".

Coming to the switch,

$$V_O = D \times V_{IN} \quad \textit{(buck)}$$

Therefore, differentiating

$$\frac{dV_0}{dD} = V_{IN}$$

So, in very simple terms, the required transfer function of the intermediate "duty-cycle-to-output stage" (i.e. the switch) is equal to V_{IN} for a buck. And it is not frequency-dependent either. Just a DC block.

All the frequency dependent response of the plant comes from its LC post filter.

Finally, the control-to-output (plant) transfer function is the product of the three (cascaded) transfer functions, i.e. it becomes, using $s = j\omega$, and $j = \sqrt{(-1)}$:

$$G(s) = \frac{1}{V_{RAMP}} \times V_{IN} \times \frac{1/LC}{s^2 + s\left(1/RC\right) + 1/LC} \qquad \textit{(buck: plant transfer function)}$$

The LC post filter (third term) above will be discussed in more detail later. Here L is the buck inductor, C the output capacitor, and R the load resistor across the output terminals of the buck. This is an approximation so far, because we are ignoring the ESR (equivalent series resistance) of the output capacitor, and the DCR (DC resistance) of the inductor. Alternatively, this simplified plant transfer function can be written out as:

$$G(s) = \frac{1}{V_{RAMP}} \times V_{IN} \times \frac{1}{\left(\frac{s}{\omega_0}\right)^2 + \frac{1}{Q}\left(\frac{s}{\omega_0}\right) + 1} \qquad \textit{(buck: plant transfer function)}$$

where $\omega_0 = 1/\sqrt{(LC)}$ is the resonant (break) frequency of the LC post filter, and $\omega_0 Q = R/L$. Or equivalently, $Q = R\sqrt{(C/L)}$.

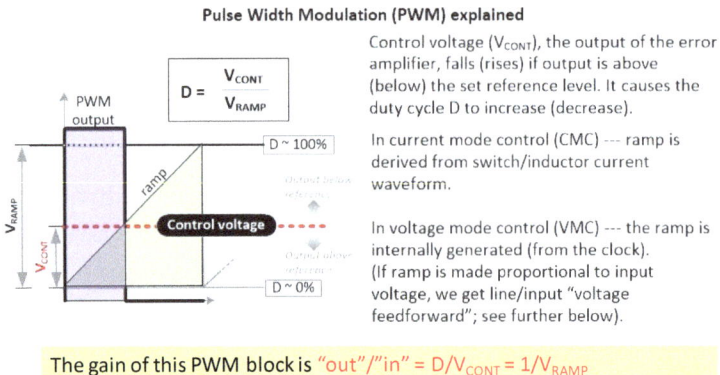

Pulse Width Modulation (PWM) explained

Control voltage (V_{CONT}), the output of the error amplifier, falls (rises) if output is above (below) the set reference level. It causes the duty cycle D to increase (decrease).

In current mode control (CMC) --- ramp is derived from switch/inductor current waveform.

In voltage mode control (VMC) --- the ramp is internally generated (from the clock). (If ramp is made proportional to input voltage, we get line/input "voltage feedforward"; see further below).

The gain of this PWM block is "out"/"in" = D/V_{CONT} = $1/V_{RAMP}$

Note that gain of any block need not be of the form voltage/voltage. Here, for example, it is duty cycle/control voltage.

FIGURE 17: GAIN OF PWM STAGE

A way to speed up the response of a buck or buck-derived switcher (such as a forward converter) is to introduce a way of sensing the input voltage and changing its slope *proportionally*, as shown in **Figure 18**. So if the input increases by a certain factor, the duty cycle automatically decreases by the same factor—as is desirable based on the simple buck duty cycle equation $D = V_{OUT}/V_{IN}$, which basically implies that for a given output, the duty cycle is inversely proportional to the input.

With this feedforward technique, there is now "input/line rejection" because the correction to the line transient (any frequency or rate) is almost instantaneous. It barely depends on the control loop to act (belatedly), through all its inherent resistor-capacitor combinations and other delays. Indeed, the control loop will eventually fine-tune the correction, but the bulk of the correction is already complete, through this feature.

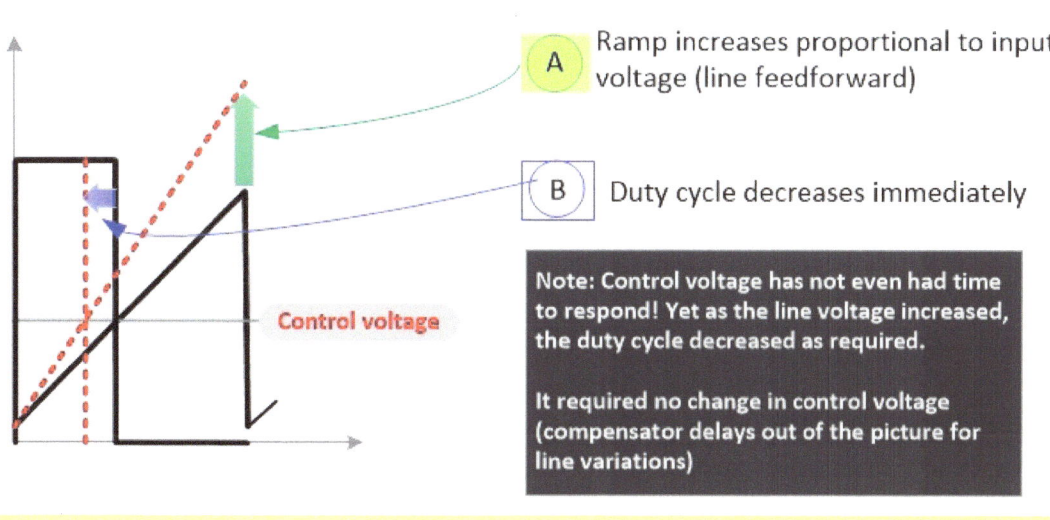

Line Feedforward explained

A — Ramp increases proportional to input voltage (line feedforward)

B — Duty cycle decreases immediately

Control voltage

Note: Control voltage has not even had time to respond! Yet as the line voltage increased, the duty cycle decreased as required.

It required no change in control voltage (compensator delays out of the picture for line variations)

VMC with line feedforward is nowadays considered superior to CMC

FIGURE 18: INPUT VOLTAGE FEEDFORWARD

One of the oft-touted historical advantages of CMC, is *inherent line rejection*. Let us look at that a bit.

Figure 17 implies that the PWM ramp is created artificially from the fixed internal clock of the switcher. That is voltage mode control (VMC) of course. In current mode control (CMC), the PWM ramp is an appropriately amplified version of the switch/inductor current.

Though the line feedforward technique described in **Figure 18** is applicable only to VMC, the original inspiration behind the idea does come from *current mode control*— in which the PWM ramp, generated from the inductor current, automatically increases if the line voltage increases. That partly explains why current mode control seemed to respond so much "faster" to line disturbances than traditional voltage mode control at the time.

However, once **Figure 18** has been implemented, VMC has effectively imbibed the key advantage of CMC. One question remains: how good is the "built-in" automatic line feedforward of CMC, compared to VMC with line feedforward? *It turns out the latter is better*. Because in a buck topology, the *slope* of the inductor current up-ramp is equal to $(V_{IN}-V_O)/L$. So if we double the input voltage, we do *not* end up doubling the slope of the inductor current or the PWM ramp as desired. That means the duty cycle does not halve exactly, as we want it to, based on $D = V_{OUT}/V_{IN}$. However, in the case of VMC with line feedfoward, it does exactly that as explained in **Figure 18**.

In other words, voltage mode control with proportional line feedforward control, though inspired by current-mode control, provides *better* line rejection than current mode control (for a buck).

For a boost topology, using its duty cycle equation, we can similarly derive the gain of the switch stage.

$$V_O = \frac{V_{IN}}{1-D}$$

$$\frac{dV_O}{dD} = \frac{V_{IN}}{(1-D)^2}$$

So the plant transfer function is a product of three transfer functions:

$$G(s) = \frac{1}{V_{RAMP}} \times \frac{V_{IN}}{(1-D)^2} \times \frac{\frac{1}{LC} \times \left(1 - s\left(\frac{L}{R}\right)\right)}{s^2 + s\left(\frac{1}{RC}\right) + \frac{1}{LC}} \qquad \text{(boost: plant transfer function)}$$

where **L** = **L/(1-D)²** as discussed previously. It is *the inductor in the "equivalent post-LC filter"* *of the canonical model.* Also note that C remains unchanged. It is just C_{OUT}.

Alternatively, the above transfer function can be written as

$$G(s) = \frac{1}{V_{RAMP}} \times \frac{V_{IN}}{(1-D)^2} \times \frac{\left(1 - \frac{s}{\omega_{RHP}}\right)}{\left(\frac{s}{\omega_0}\right)^2 + \frac{s}{\omega_0 Q} + 1} \qquad \text{(boost: plant transfer function)}$$

where $\omega_0 = 1/\sqrt{(LC)}$, and $\omega_0 Q = R/L$.

We have included a surprise term in the numerator, the RHP zero, which can be shown to be present in both the boost and buck-boost, after detailed modeling. Its location is

$$f_{RHP} = \frac{R \times (1-D)^2}{2\pi L} \qquad \text{(boost)}$$

Similarly for a buck-boost we get:

$$V_O = \frac{V_{IN} \times D}{1-D}$$

$$\frac{dV_O}{dD} = \frac{V_{IN}}{(1-D)^2}$$

(Yes, it is an interesting coincidence --- the slope of 1/(1-D) calculated for the boost, is the same as the slope of D/(1-D)) calculated for the buck-boost!)

So the control-to-output transfer function is

$$G(s) = \frac{1}{V_{RAMP}} \times \frac{V_{IN}}{(1-D)^2} \times \frac{\frac{1}{\underline{L}C} \times \left(1 - s\left(\frac{\underline{L}D}{R}\right)\right)}{s^2 + s\left(\frac{1}{RC}\right) + \frac{1}{\underline{L}C}} \qquad \textit{(buck-boost: plant transfer function)}$$

where $\underline{L} = \textbf{L}/(\textbf{1-D})^2$ is the inductor in the *equivalent* post-LC filter.

Alternatively, this can be written as

$$G(s) = \frac{1}{V_{RAMP}} \times \frac{V_{IN}}{(1-D)^2} \times \frac{\left(1 - \frac{s}{\omega_{RHP}}\right)}{\left(\frac{s}{\omega_0}\right)^2 + \frac{s}{\omega_0 Q} + 1} \qquad \textit{(buck-boost: plant transfer function)}$$

where $\omega_0 = 1/\sqrt{(\underline{L}C)}$, and $\omega_0 Q = R/\underline{L}$.

Note that, as for the boost, we have included the RHP zero term in the numerator (in gray). Its location is similarly calculated to be

$$f_{RHP} = \frac{R \times (1-D)^2}{2\pi L \times D} \quad \textit{(buck-boost)}$$

ENTER: THE LAPLACE TRANSFORM

Some of the rather unexpected situations described above, and others we will encounter, can be equally unexpectedly visualized more elegantly, in terms of the "dreaded" Laplace transform technique. We need to feel comfortable with it, and for that reason we will discuss it a bit here. We will also take this opportunity to restate, summarize or emphasize some of our key "lessons learned".

The best way to quell our fear of the Laplace transform is to understand that we are simply moving to an alternative *mathematical* domain to simplify computation. We have been doing the same thing for years using log math. See **Figure 19** and **Figure 20**. Complex multiplications or divisions of very large numbers get reduced to easy addition and subtraction instead. Of course, we rely on tables previously created, to go in and out of this logarithmic plane. So in a sense, the spade work was already done once and for all, by creating the log and antilog tables. Because, to return to the normal, non-logarithmic plane, we have to use antilog or inverse-log tables.

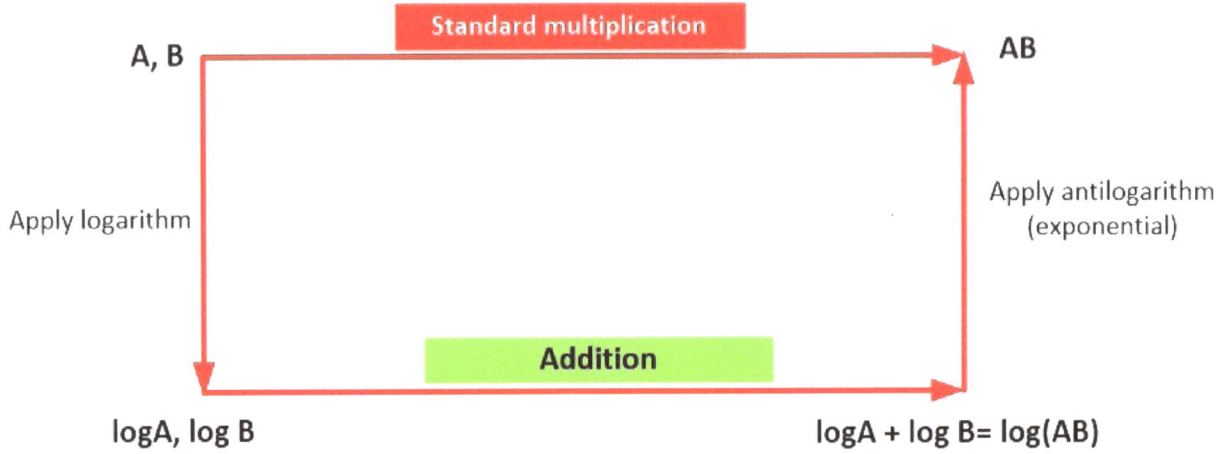

FIGURE 19: USING LOGARITHMS TO SIMPLIFY MULTIPLICATION OF LARGE NUMBER

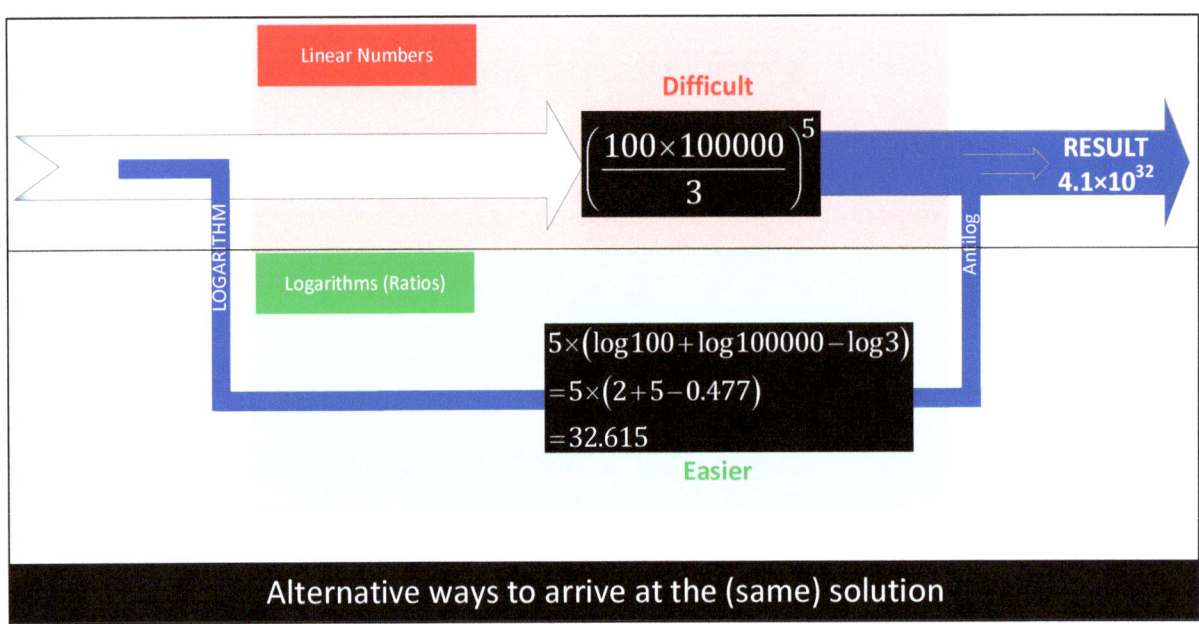

FIGURE 20: VISUALIZING THE LOGARITHMIC PLANE

See **Figure 21** now, for the Laplace transform technique. This figure does *not* intend to show a closed-loop control system. Think of it for example, as just a simple filter stage, consisting of various capacitors, resistors and inductors. We apply an arbitrary input signal or impulse, and we are interested in seeing what happens at the output of this network. We discover that the differential equations to solve this problem (in the normal "time domain") become very complicated.

It also turns out that using the alternative "frequency domain" or "s-plane", i.e. the Laplace transform, the math becomes simpler. But once again we rely on a bunch of readily available tables, with the help of which we can move in and out of the new *mathematical computation domain*.

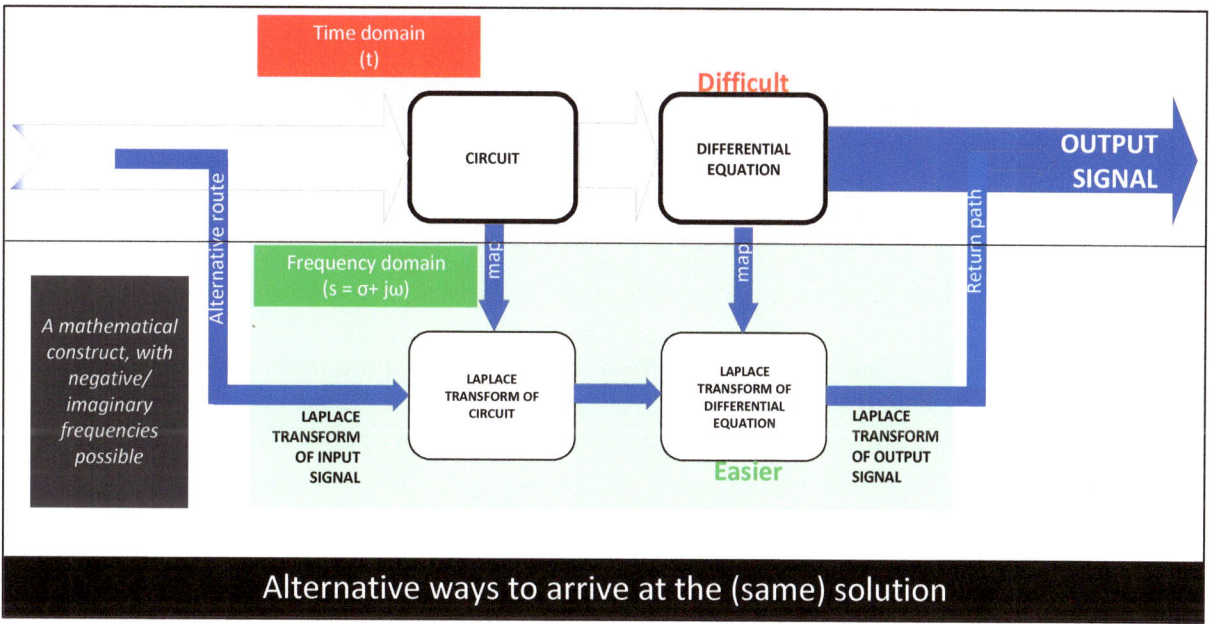

FIGURE 21: THE LAPLACE TRANSFORM TECHNIQUE (FREQUENCY DOMAIN ANALYSIS)

What exactly are we achieving by the Laplace transform? Essentially, we are breaking up an arbitrary non-repetitive, time-varying signal or impulse (the "disturbance"), into a *continuous* spectrum of both positive and negative frequency components (i.e. in the frequency domain). This is akin to the well-known Fourier analysis technique used for decomposing *repetitive* waveforms into *discrete* (positive) harmonics. Note that decomposition approaches are easier for the same reason that we routinely break up "force" in classical mechanics, into its "independent" x, y and z components, do the math for each "independent" axis separately, compute the x, y and z components of the acceleration, and finally sum them up vectorially to give us the final acceleration vector: i.e. its magnitude and direction. That is what we are doing in the Laplace transform method too, quite similar to what we did in high-school with the Fourier series, except that the decomposed frequencies are now a continuous spectrum. We will come to this in a little more detail in the next chapter.

Summarizing, think of **Figure 21** as a simple 2-port network, say a combination of several resistors and capacitors, with an input time-varying excitation (voltage or current), and we are trying to deduce the output waveform. Usually we will need to set up complicated differential equations to solve the problem. However, the Laplace transform method allows us to take the Laplace transform of *both the circuit and the excitation*. As a result, this becomes a simple algebraic problem where we can sum over (integrate in our case) the result of the various frequency components. Finally, we take the *inverse* Laplace transform to map the result back into the time domain. We thus see the desired output voltage (or current). As mentioned, the reason we get away with this "simplification" is that all the drudgery has already been done beforehand in the form of comprehensive lookup tables for Laplace and inverse Laplace transforms.

With the Laplace transform technique, we can show that a certain time delay is equivalent to a "phase lag" in the frequency domain, one which is proportional to the frequency of the component. See: http://lpsa.swarthmore.edu/BackGround/TimeDelay/TimeDelay.html

As previously mentioned, in switching power supplies too, data is not sampled and acted upon continuously. In other words, there are inherent delays on account of the stream of discrete pulses. That also implies there will be increasing phase lag as we approach frequencies close to the switching frequency. To avoid meeting our criterion of instability (T = -1) prematurely, we must start by fixing the crossover frequency to less than one-fifth the switching frequency.

That makes sense because the term "phase" (angle) has no relevance unless we are talking in terms of a specific frequency and its associated time period (of repetition)—which we are doing in this case, through our frequency decomposition technique via the Laplace transform.

The "compensator" (feedback network) of the control loop can also introduce additional delays, with corresponding frequency-dependent phase shifts, which add to the inherent delays present in the response of the plant to the disturbance. These compensator delays are easier to understand, because the feedback circuit typically involves several resistors and capacitors, with a bunch of interrelated RC time constants. In particular, the capacitors present may also need a finite time to charge or discharge to their new average values through all their accompanying resistors. And that incidentally leads to the compensator's "poles" or "zeros", depending on how the elements are arranged within the high-gain error amplifier of the compensator. We will talk of Type 2 and Type 3 analog compensators in more detail soon.

We have seen that mathematically, if T = -1, we get full-blown instability. Intuitively, we can visualize instability as a situation where the system is trying to respond to a completely *outdated/ delayed command* without realizing it (reading "up" instead of "down" for example), and thus continues to head in the opposite (wrong) direction every time. The delay is response now is exactly one half-cycle, the word cycle referring to the decomposed component's culprit frequency. This is full-blown instability, even though the system itself does not literally explode.

As mentioned, the majority of "disturbances" or excitations encountered in switching power supplies are not "small-signal" as many probably still assume, but large-signal. Under such bulk stimuli, it can take a certain finite time for the current in an inductor to slew up or down to the new desired "average" value commensurate with the new steady load condition. See **Figure 4** once again. We also remember that this inability of the inductor and the switch to provide the required power during this "reinitialization" duration is *equivalent to a sudden drop in gain*—as the closed-loop system struggles to correct itself. This can lead to full-blown instability if "conditional stability" is already present.

The maximum slew rate of the inductor current is primarily dependent on the "timeless", and shall we say "stubborn", equation of an inductor: $V = L\, \Delta I/\Delta t$. Though there are ways to mitigate the resulting delay, such as hysteretic control, ultimately, this is the downside of using reactive (energy-storing) components (inductors and capacitors), compared to the fast-acting but wholly *inefficient* resistive elements employed in linear power supplies. We always need *time*, to either build up or deplete stored energy in a controlled manner. But we can dissipate almost immediately! That is equivalent to saying: resistors have no inherent delays.

And that is why, as mentioned earlier, if we suddenly go from 0A to 5A load in a switcher for example, the initial dip in the output voltage may have little to do with the control loop characteristics. Indeed, the control loop can always make things worse, but even if it is considered optimized here, the output response may eventually be determined only by the output bulk capacitance vis-à-vis the inductance, since the output capacitor needs to supply the entire additional energy demanded till the current in the inductor can ramp up and take over. And if the capacitance happens to be inadequate to start with, we need to refer back to **Figure 7**.

Similarly, as also explained in **Figure 7**, if we suddenly go from 5A to 0A (unloading), we may find to our horror that though we saw the output voltage jump up and respond by even *halting switching action completely*, the output rail continued to rise, almost out of our control for a short while. The reason for that is the inductor stubbornly pushes out all the stored energy related to its initial current setting, into the output capacitors where it can be stored indefinitely as electrostatic energy as required, since the attached load isn't demanding energy anymore.

Perhaps that reminds us of the hazy outlines of our forgotten Physics 101 course: *energy can be converted* or dissipated as heat (in resistors), *but never wished away*. That is why we need catch diodes (and output capacitors) in any switcher in the first place. To freewheel the current associated with the stored magnetic energy. Dispense with the diode, and we get a worthy spark ignition system instead of a switcher—lots of heat and light, but no useful power.

But there are still some remaining "subtleties" with regards to *non-buck* topologies, as shown in **Figure 4**. For example, during a line (input) disturbance, as opposed to a load transient, things become quite different for the boost and the buck-boost topologies, as compared to the simple scenario described in the figure for a buck topology on the left-hand side. The reason is, in a buck, the average inductor current equals the load current (in steady state), and is therefore constant during a line disturbance. There is no delay attributable to any inductor "reinitialization" problem, except of course for a load transient.

However, in a boost or buck-boost, the average inductor current *is* a function of the duty cycle, unlike a buck, and thus needs to move to a new average value if we vary the input voltage, even if the load current is held constant! So now, the inductor reinitialization issue returns to haunt us, along with the other inherent delays present in the control loop—even during a supposedly "pure line disturbance"! See the right-hand side of the figure.

This goes to show that not all "disturbances" are alike, *nor all topologies*, and we must be cognizant of quite a few such unexpected "subtleties" when we try to move basic control theory concepts over to switchers.

We remind ourselves once again that to prevent smaller errors of perception from snowballing into masses of confusion, we must be very clear about the exact meaning of all the terms in common use.

As mentioned, a prime example of that is the concept of "closed loop gain". And the other side of the very same silver coin: "(open) loop gain" or "T". Many power supply engineers continue to think that open loop gain is some sort of amplification factor that gets applied to a vague, unspecified "disturbance" when the feedback loop is *literally* "open": i.e. broken or non-existent. Then, as a corollary, they assume that in contrast, closed loop gain must be what we measure when the feedback loop is actually present! A sideshow of this confusion is that a hands-on power supply engineer may wonder why, when he or she runs Bode plots using a standard bench network analyzer, the machine claims that it is measuring the "open loop gain". "Why isn't it giving us the closed loop gain, considering the fact that the loop is in reality closed?" And so on. One wrong premise leads to many wrong conclusions. Some engineers wisely use the term "loop gain" instead of open loop gain, as we too eventually finally did on previous pages. Others prefer to call "T" the "round transfer function". It can get a bit confusing.

Many switcher engineers/authors did realize early on that closed loop gain was V_{OUT}/V_{REF}, not V_{OUT}/V_{IN}. But then, instead of giving examples showing line/load disturbances, they inadvertently propagated the fallacy further by documenting the overshoots and ringing when the reference voltage is ramped up suddenly from 0V. For example, we will often see in related literature the case of a "1/s" "step disturbance" applied to the system, where s = jω as usual. But in this case, the step is the *shape of the reference voltage*. You may wonder why it is relevant.

We need to remember that:

a) Every power converter starts up initially with its reference rising up to its set value, so that hardly qualifies as a "disturbance" of interest to us.

b) Besides, the reference voltage, and the output voltage, rarely come up "instantly", since in a practical case both are usually brought up gradually under the influence of a closed-loop soft-start circuit. The reference is typically slow to rise, as it comes via a 0.1 μF ceramic decoupling capacitor placed on the current-limited REF pin, which is charged up slowly.

c) Even if we assume V_{REF} did come up abruptly, the output itself would take a very long time after that, relatively speaking, to ramp up to its steady value, since it has to first charge the rather sizeable output bulk capacitance across it, through the intervening (slew-rate limiting) inductor. So this response scenario has really nothing to do with jerking/wiggling the reference voltage around, even if that is considered relevant.

Indeed, the way the output comes up *with no soft-start implemented*, may show some ringing which *qualitatively* mimics the ringing observed during line and load transients.

We should also point out that a practical, often overlooked problem to measuring what some engineers still call "open loop gain", is that in a modern high-gain switcher, there may be no easy way to "break" up the loop, i.e. to literally open it, without causing disastrous effects on the output. Leave aside testing it successfully in that state.

Indeed, that can be done on occasion. As when trying to stabilize relatively *low-gain* mag-amp post-regulators, using the oft-mentioned K-factor method (see http://www.ti.com/lit/ml/slup129/slup129.pdf).

The venerable K-factor method from Mr. Venable, a subject of many popular articles on feedback control, implicitly requires knowledge of the gain with *no* feedback present, i.e. with the loop broken—as a means of optimizing the feedback loop when it is finally introduced. And so, even though some engineers insist that the K-factor is all that is ever required for stabilizing switchers, it is usually impractical in most modern cases.

Besides the practical aspects, the K-factor technique also unfortunately trades off gain for phase margin by reducing gain at low frequencies and increasing it at higher frequencies—quite the opposite of what we usually try to do, for reasons we will soon discuss. That is why the K-factor technique is perhaps only well-suited for post-regulators, where a steady almost ripple-free DC input rail is present to start with— such as mag-amps! But rarely otherwise.

The K-factor method as applied to a Type 3 amplifier attempts to put two coincident zeros a factor of $1/\sqrt{K}$ below crossover and two coincident poles a factor of $\sqrt{K}$ above crossover. As we will see, that does nothing to the *peaking of the LC pole*, which is not only a potential cause of conditional stability, but affects the ringing at the output during line/load transients as we will soon uncover.

But the irony of it all is, perhaps all the effort that the K-factor based "optimization" effort represented, was ultimately for only a questionable improvement in something called the phase margin. Questionable, because no one seems to agree fully what is the "optimum phase margin", *and why*.

Let's recapitulate: stability ultimately depends on the following basic question: What happens if the disturbance undergoes various delays as it goes around the loop? For example, even in the simple case of a thermostatic room air-conditioner, a) it may take time for the sensor to feel the change if a window is thrown open. b) After that, the heater or air-conditioner will need some time to activate and respond. And so on. As also mentioned earlier, these delays can be modeled as frequency-dependent phase lags. And so, though we have usually been writing out the gain functions as simply T, G, H, etc., in reality they should be written as T(s), G(s), H(s), etc., indicating their *frequency dependence* and *inherent phase angle* too, besides their magnitude.

Similarly, for a switcher, we can now visualize a situation where an *additional 180° phase shift* can easily occur for some unspecified "harmonic" (frequency component) of the original disturbance. It will then reinforce itself after going around the loop. The result is the system could break up into oscillations.

For rejecting low- to mid-frequencies, it is now obvious that we need to try and maximize the DC gain. But we also need to deliberately *roll the gain off at higher frequencies* to avoid instability. Ultimately, we have to adhere to a simple criterion: *We have to ensure that at the specific frequency where an additional 180° of phase lag occurs, the gain falls below 1 to avoid oscillations (gain margin).* That will ensure any disturbance will get abated as it goes around the loop. Alternatively, we must ensure that when the signal comes around full-circle with a gain of 1, its phase lag is not enough to reach the ominous level of -180°.

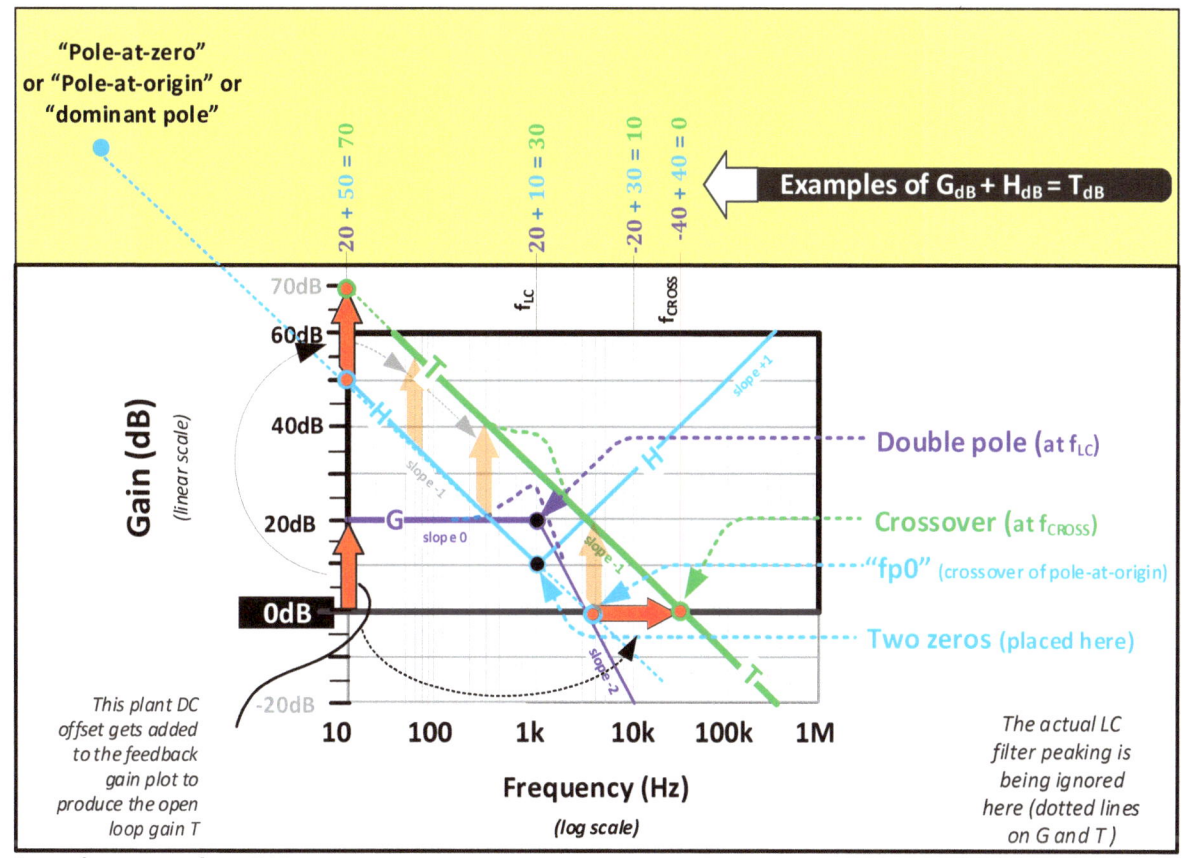

Ignoring capacitor ESR-zero

FIGURE 22: CLASSIC ANALOG CONTROL LOOP DESIGN (SIMPLIFIED) AND RESULTING LOOP GAIN

In **Figure 22** we show a typical compensation exercise, but in terms of gain magnitudes only so

far. It breaks up the "almost" straight line of T into its constituent G and H components. Note

that the double LC "pole" of the plant, which is responsible for the "-2" slope of G after the

breakpoint, has been (roughly) canceled out by placing *two "zeros" at the exact position of the*

LC, by the compensator. So we are left with an (almost) straight line for T, essentially coming

from the compensator's low-frequency gain profile. Though that is displaced vertically by the

exact amount equal to the DC gain of the plant, as we will see shortly.

Note that the loop gain T is simply the product of the two cascaded gains, G and H, but on a log plane, things are easier. We can just sum the two curves of G_{dB} and H_{dB}, to give us T_{dB}. In other words, once we express the gains in decibels, we can just sum them up, rather than multiply them out, as shown in the figure.

In the figure, we can also see we have set a very high DC gain (via the compensator) —as recommended for attenuating low- to mid-frequency disturbances. In fact, theoretically, the gain is infinite at 0 Hz, but in reality it gets limited by the inherent characteristics of the error amplifier, though that is not shown in the figure for the sake of simplicity, and also because it is practically impossible to display 0 Hz on a log scale anyway. But we have shown a dotted line extending to some very low frequency, and that is called the "pole-at-origin" or "pole-at-zero", among other monikers. But it has no location that we can really specify or draw out. What we do know however is, wherever it is located, it causes the gain to fall at "-1" slope thereafter. So its exact location, i.e. how low-frequency it really is, is reflected only by *the frequency at which it intersects the unity gain (0dB) axis*. That frequency is what we are calling "fp0" here. Indeed, we may place two zeros in the compensator well before the H_{dB} curve ever gets to cross the 0dB axis. But if we draw a dotted line to the 0dB axis, the intersection frequency is fp0 as shown in **Figure 22**.

"fp0", the crossover frequency of the pole-at-origin, needs to be set very carefully because it is the key parameter which ultimately determines the crossover frequency of interest to us: f_{CROSS} (the crossover of T). Both fp0 and f_{CROSS} are related through the DC gain of the plant, which in turn is completely responsible for the vertical arrow shifts shown in **Figure 22**. We will now see what that exact relationship is.

The plant, as we know by now, has three main constituents, and its gain is the product of its three stages.

$$G(s) = \frac{1}{V_{RAMP}} \times V_{IN} \times \frac{1/LC}{s^2 + s\left(1/RC\right) + 1/LC}$$

So its DC gain (the flat portion of G up to the break frequency in **Figure 22**) is simply V_{IN}/V_{RAMP} (or 20 log of that in decibels). As we can also see from this figure, this is the amount by which the compensator gain profile gets shifted upwards to create "T". Since we are dealing with a "-1" (inverse proportionality) curve for T, it is easy to see that the following relationship tells us how exactly we must position fp0 from the compensator, to achieve a certain desired f_{CROSS} for T.

$$fp0 = \frac{V_{RAMP}}{V_{IN}} \times f_{CROSS}$$

So, the loop gain curve in **Figure 22** "crosses over" at a frequency f_{CROSS}, which implies a gain of 1 at that frequency (on a log scale it is the zero of the y-axis since log 1 = 0). To stay well away from any phase lag effects causing total instability on account of the discrete/sampling issues related to the switching frequency of switchers, f_{SW}, it is customary to set f_{CROSS} to at least less than $f_{SW}/2$ ("Nyquist's sampling criterion"). But in fact, it is far better to set the crossover somewhere between $f_{SW}/10$ to $f_{SW}/5$. Not higher.

Note that the 180° inherent phase lag on account of negative feedback ("negative" though only at low frequencies as we now realize), is rarely plotted out. It is "understood". Only the *additional* phase shift introduced by the feedback network and the plant combined, is displayed on a typical Bode plot.

There are also other parasitics that come into the picture, which we have neglected so far. One is the ESR-zero, coming from the ESR of C_{OUT}. Using a Type 3 compensator, we try to kill this zero (from the plant), by placing a pole at the same exact position. But besides the pole-at-origin, a Type 3 compensator produces 2 zeros (both of which we have used up already, to cancel the LC double pole), but also 2 poles, one of which we can use to kill the ESR-zero. See **Figure 23**. That leaves us with one additional pole to play with, called "fp2" here. Some people say we should place it at $10 \times f_{CROSS}$, others say that to attenuate the high-frequency ripple component, we need to place it at $f_{SW}/2$. Lloyd Dixon suggested placing it at exactly f_{CROSS}. This was called the "optimized solution" in this author's A-Z/2e text book.

In **Figure 24**, we have shown the poles and zeros from a typical analog control loop exercise, extracted from one of this author's A-Z/2e book. Note that this is no longer the "asymptotic approximation" used in **Figure 22**. It has all the curved regions, plotted out using Mathcad.

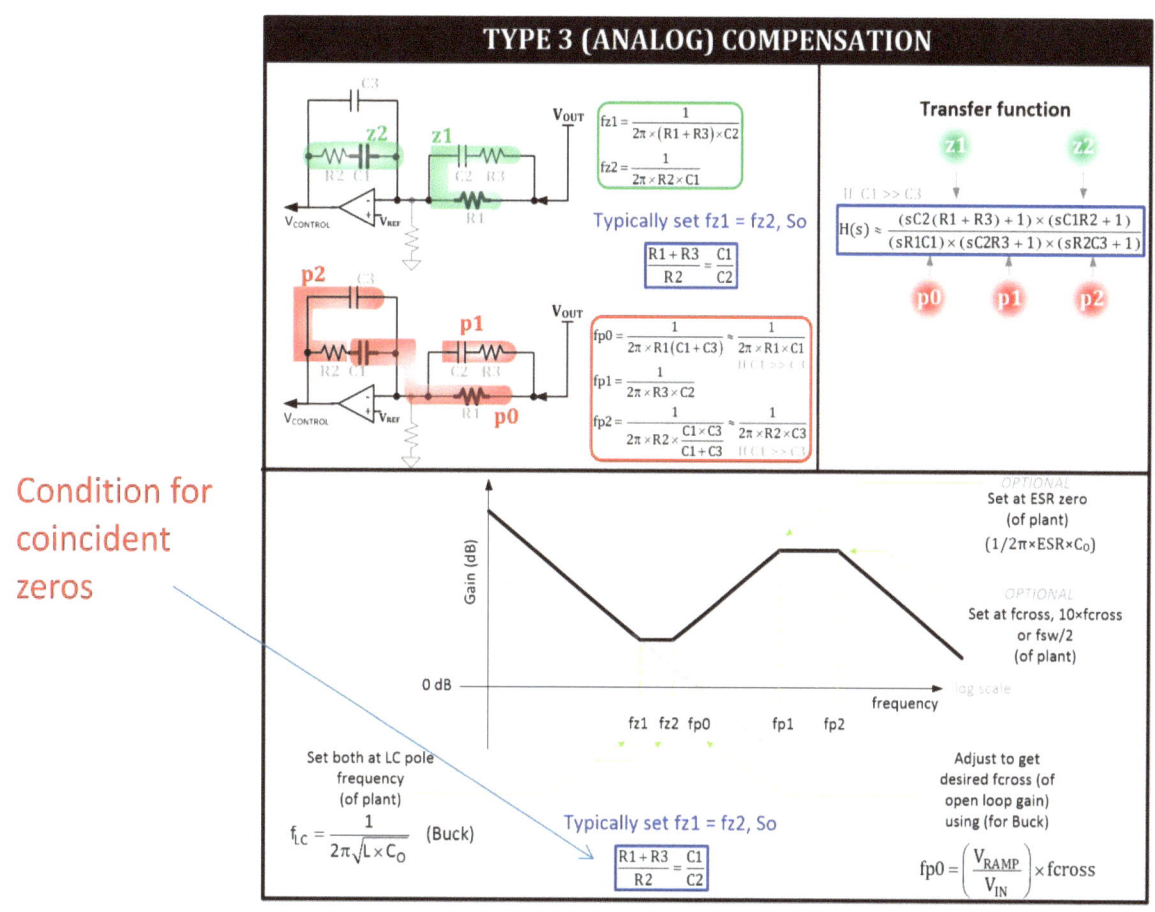

Condition for coincident zeros

FIGURE 23: TYPE 3 COMPENSATOR EQUATIONS AND STRATEGY

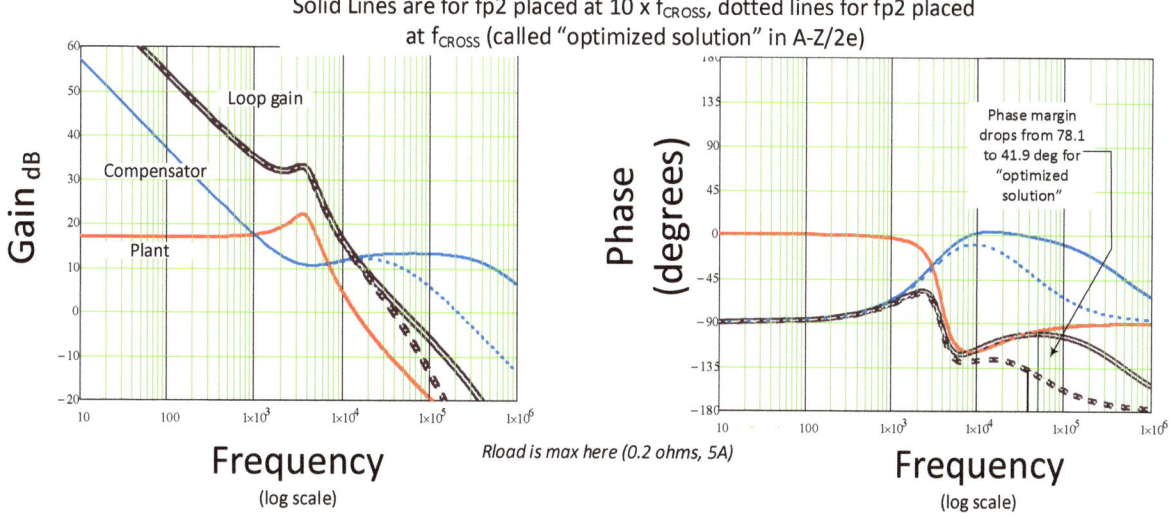

FIGURE 24: RESULTS OF AN ACTUAL TYPE 3 COMPENSATION EXERCISE AT MAXIMUM LOAD

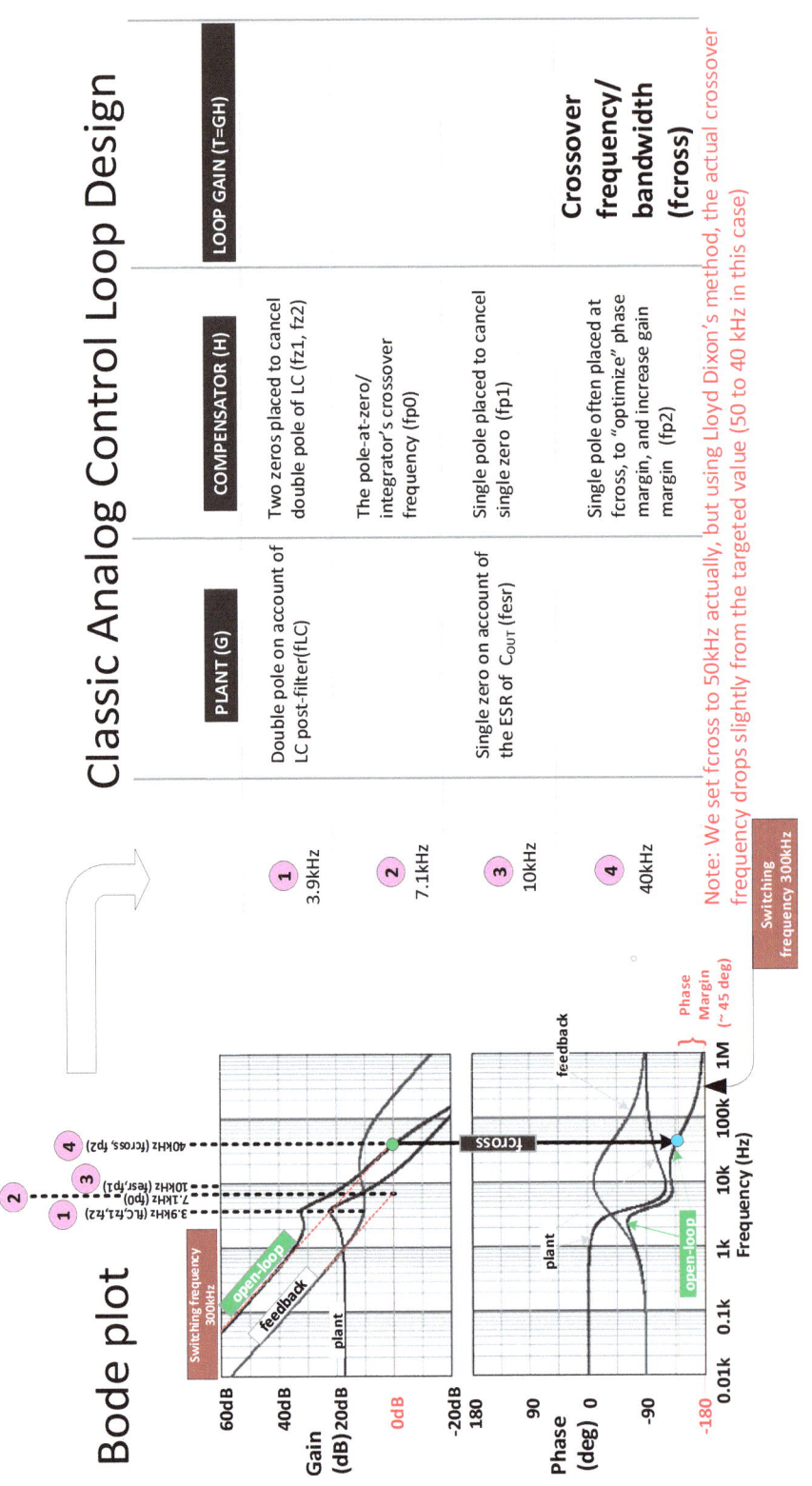

FIGURE 25: SUMMARY OF POLE-ZERO PLACEMENTS FOR PREVIOUS FIGURE AND EXAMPLE

74

In **Figure 25**, we summarize the compensation strategy we have been talking about, showing exactly what happens (in terms of the frequencies involved), for the plant G, compensator H, and of course "T".

In **Figure 26**, we present a table of the components of a Type 3 compensator, based on the transfer function equation in **Figure 23**, showing how all but one component, are involved in more than one pole/zero position. *Which is why it becomes so difficult to change anything in an analog loop.* Changing just one component can have a domino effect on the gain curves, with "unintended" consequences. We will discuss this in more detail shortly.

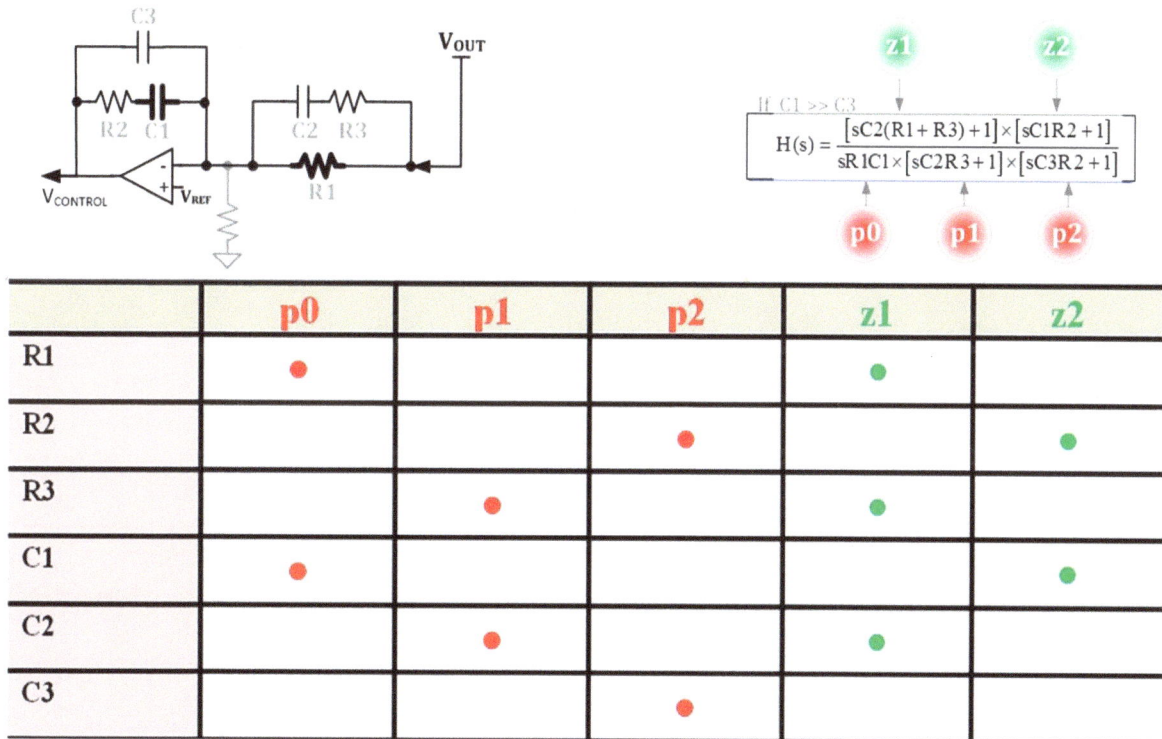

$$H(s) = \frac{[sC2(R1+R3)+1] \times [sC1R2+1]}{sR1C1 \times [sC2R3+1] \times [sC3R2+1]}$$

	p0	p1	p2	z1	z2
R1	●			●	
R2			●		●
R3		●		●	
C1	●				●
C2		●		●	
C3			●		

C3 is the only component which is connected to only one pole/zero, fp2 in this case ("the buck stops here")

FIGURE 26: HOW EACH COMPONENT OF A TYPE 3 COMPENSATOR PLAYS A MULTI-ROLE

In **Figure 27**, we show other compensators too for completeness sake. Type 2 for example, offers only one pole and one zero (besides the pole-at-origin, which all three offer). It is therefore unsuitable for VMC since we need two zeros to cancel the LC double-pole. However, some older switchers try to use the ESR-zero for that purpose, along with one zero from the compensator, and therefore do not attempt to kill the ESR-zero. However that is not an optimum solution at all. Generally, Type 2 compensators can only be used with CMC, because its plant does not have the LC double-pole, instead featuring an RC-based (first order) "load pole" as shown in **Figure 28**.

The Type 1 compensator is of no practical use *on its own*, but is a key building block of Type 2 and Type 3 compensators. It is an integrator. It is the where the pole-at-origin, fp0, comes from. The key general function behind it is plotted out in **Figure 29**. It has the form

$$H(s) = \frac{A}{s} \equiv \frac{1}{s/A} \equiv \frac{1}{s/\omega_0}$$

This crosses over at $\omega_0 \equiv \omega p0 = A$. Or equivalently fp0 = A/2π. So we can adjust it (translate it upwards or downwards), by changing A.

Note: *It is always preferable to write all pole and zero functions in the form $(s/\omega_0)^x$, to avoid needless confusion about the DC gain contribution from the functions.*

The math behind the op-amp embodiment of this function, the integrator, is shown in **Figure 30**. We get the equation for fp0 as:

$$fp0 = \frac{1}{2\pi RC}$$

Note: In a Type 3 compensator (**Figure 26**), this pole-at-origin (integrator function) is created by R1C1, nor R1C3 as commonly and erroneously assumed. The reason for that is, we are working under the assumption C1>>C3; otherwise the mathematical solutions to the locations of the poles and zeros, are extremely intricate, and thus unusable.

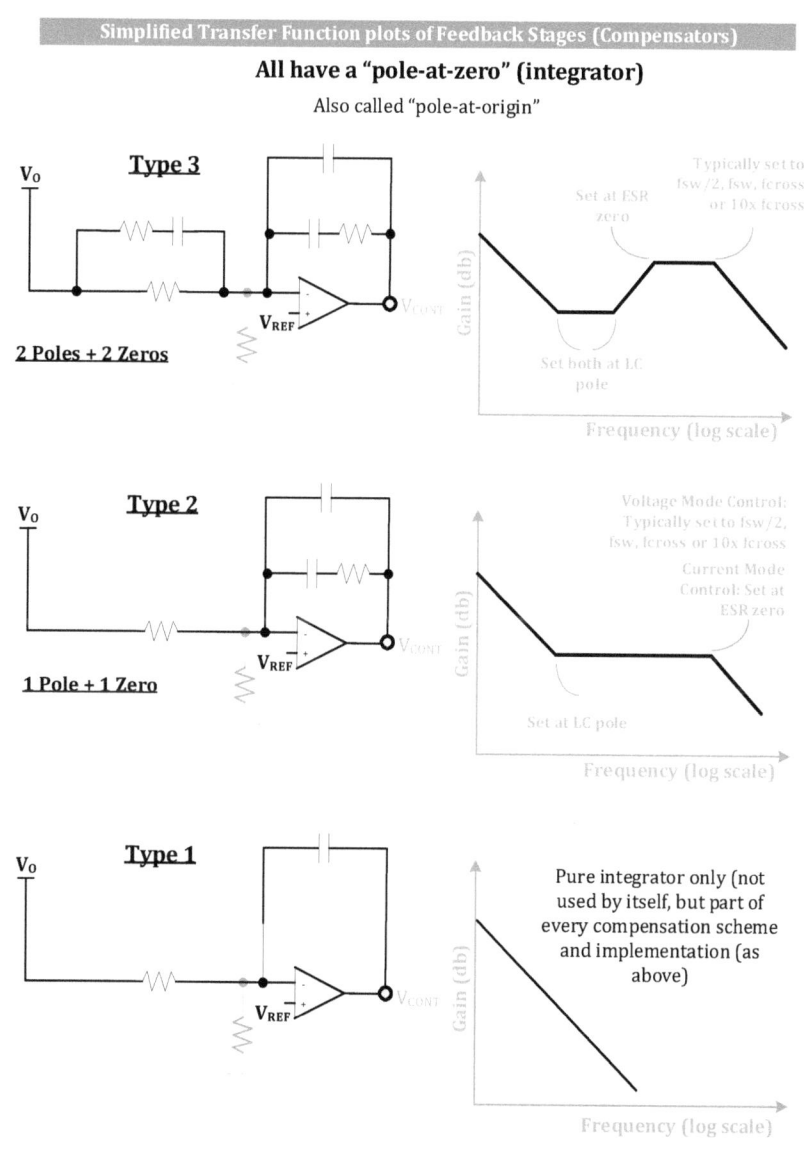

FIGURE 27: TYPE 1, 2 AND 3 COMPENSATORS

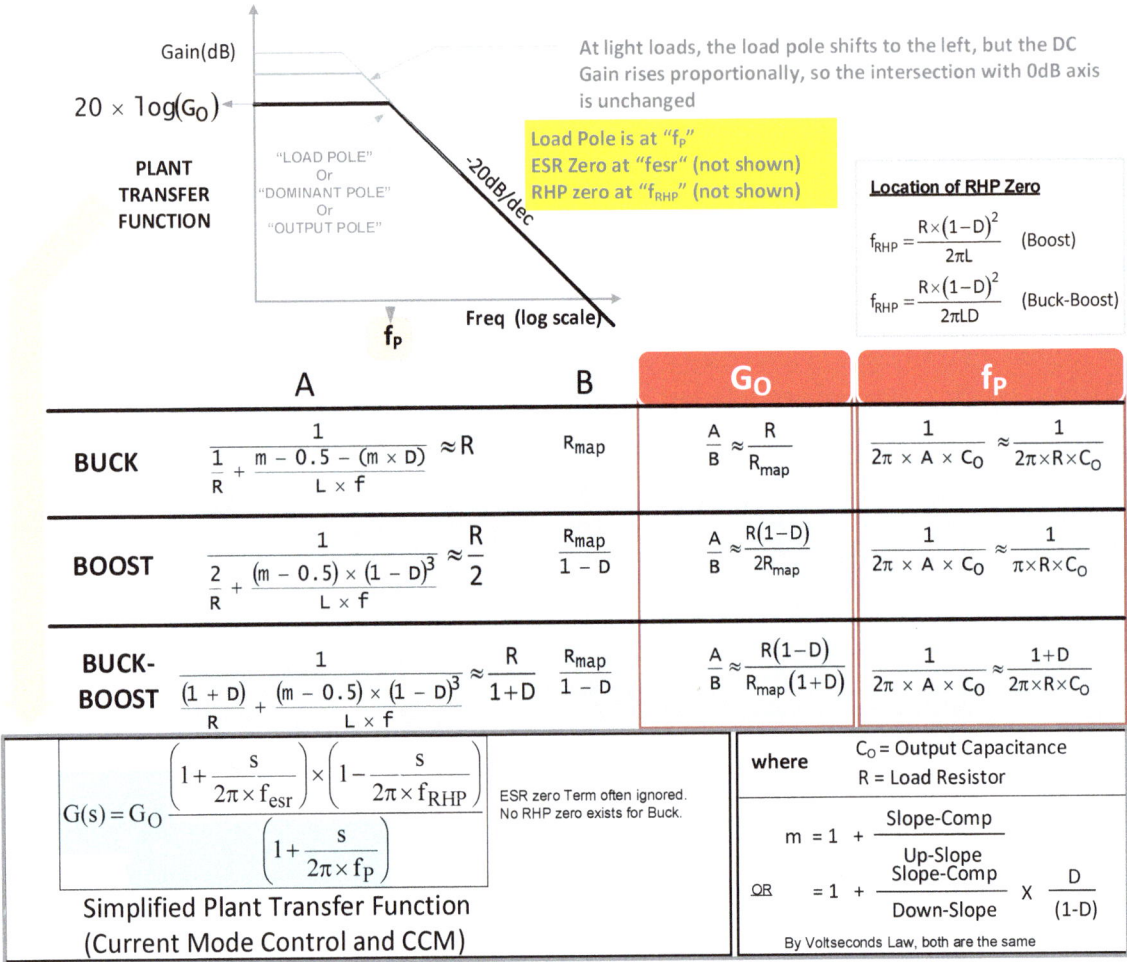

FIGURE 28: THE PLANT IN CURRENT MODE CONTROL (SUITABLE FOR TYPE 2 COMPENSATOR)

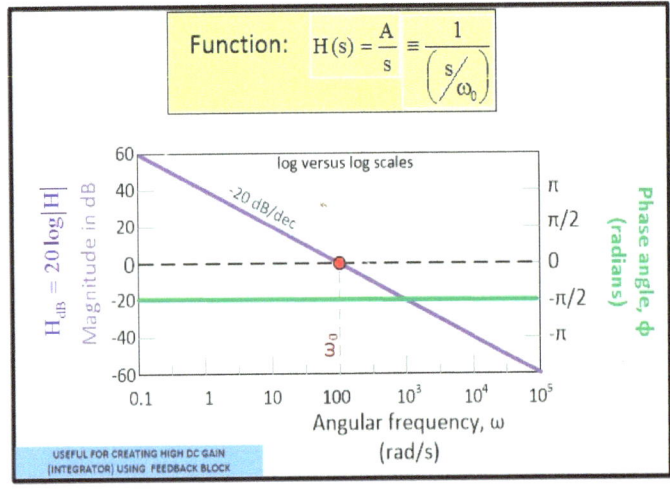

FIGURE 29: THE INTEGRATOR FUNCTION (INVERSELY PROPORTIONAL TO FREQUENCY)

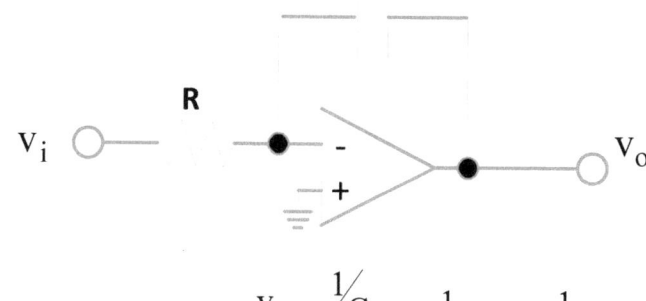

$$\text{Transfer function} = \frac{v_o}{v_i} = \frac{1/Cs}{R} = \frac{1}{RCs} \equiv \frac{1}{\left(\dfrac{s}{1/RC}\right)} \equiv \frac{1}{\left(\dfrac{s}{\omega p0}\right)}$$

So : $\boxed{fp0 = \dfrac{1}{2\pi RC}}$

FIGURE 30: ANALOG IMPLEMENTATION OF AN INTEGRATOR

As mentioned, we can luckily break up the switch and L-C combination of a boost and buck-boost into a switch followed by a distinct and equivalent cascaded post-filter stage, consisting of the same C (output capacitor), in series with an "equivalent inductor" of value:

$$L_{equivalent} \equiv \underline{L} = \frac{L}{(1-D)^2}$$

In effect, that makes the effective inductance *a function of the input voltage*. Hence the treatment can get rather complex, since the LC resonant frequency moves to higher and higher frequencies as we lower the input voltage (higher D). And that is, intuitively, what eventually contributes to the RHP zero instability mentioned earlier. The conventional solution to the RHP zero problem is to virtually accept that there is no solution! We just have to roll-off the loop gain at a much lower frequency than the typically targeted $f_{SW}/10$ to $f_{SW}/5$ for a buck. Maybe we need to go closer to $f_{SW}/20$, or even lower. Which is also why we can hardly expect excellent control loop response from, say, a typical power factor correction (boost) stage, or a "cheap and dirty" flyback (buck-boost).

Finally, we present a summary of the plant functions for VMC and CMC, for easy reference. In **Figure 31**, **Figure 32**, **Figure 33**, and **Figure 34** we have the buck, boost and buck-boost (all in VMC) respectively, followed by the buck in CMC.

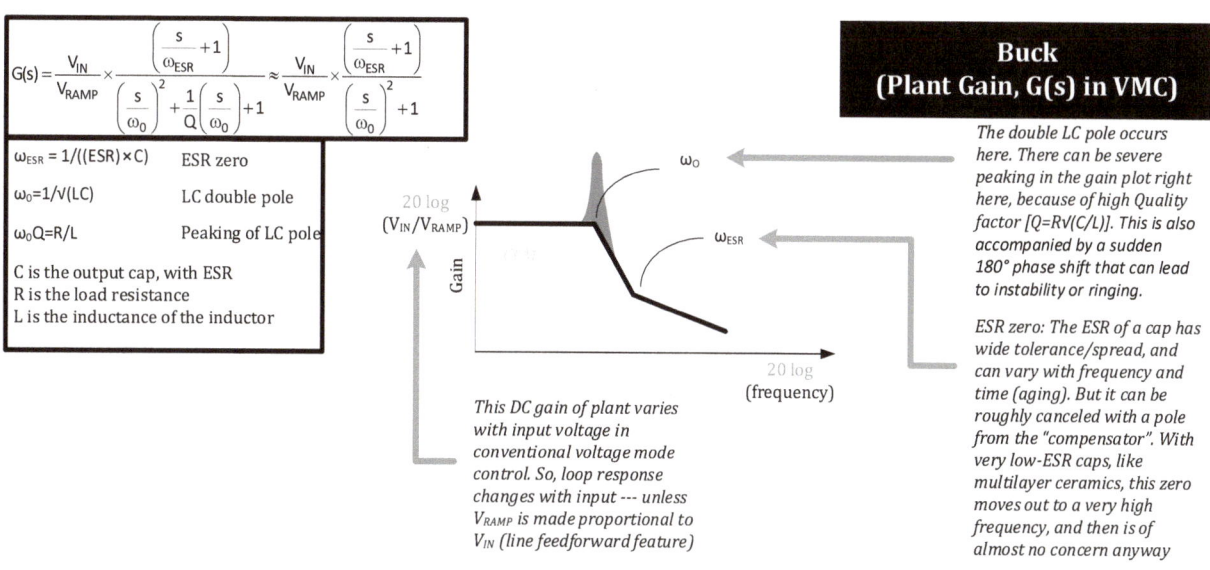

FIGURE 31: BUCK IN VMC

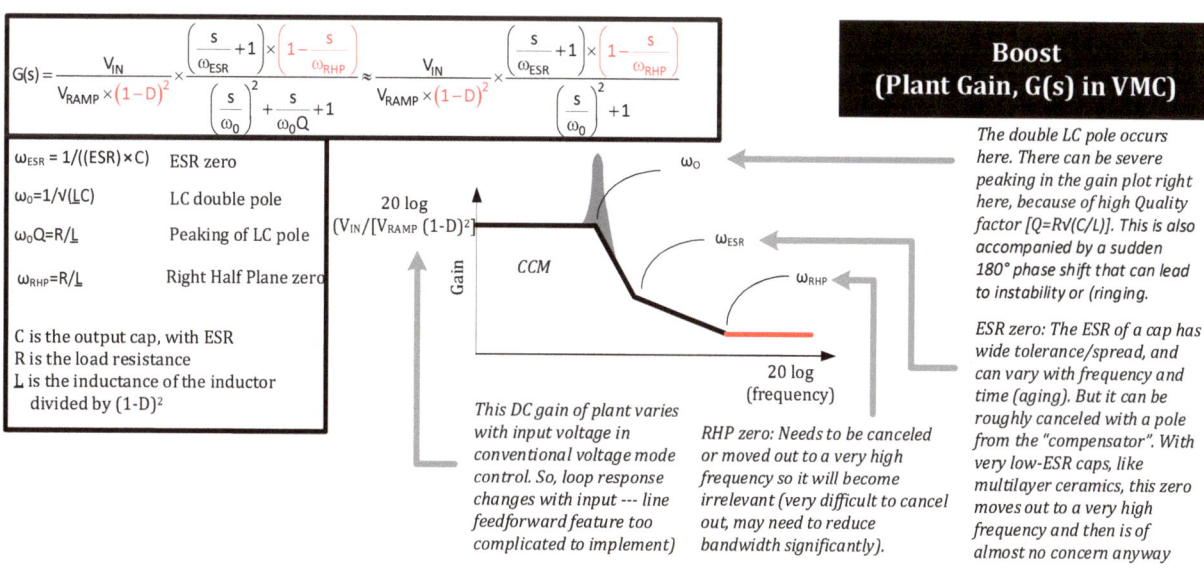

FIGURE 32: BOOST IN VMC

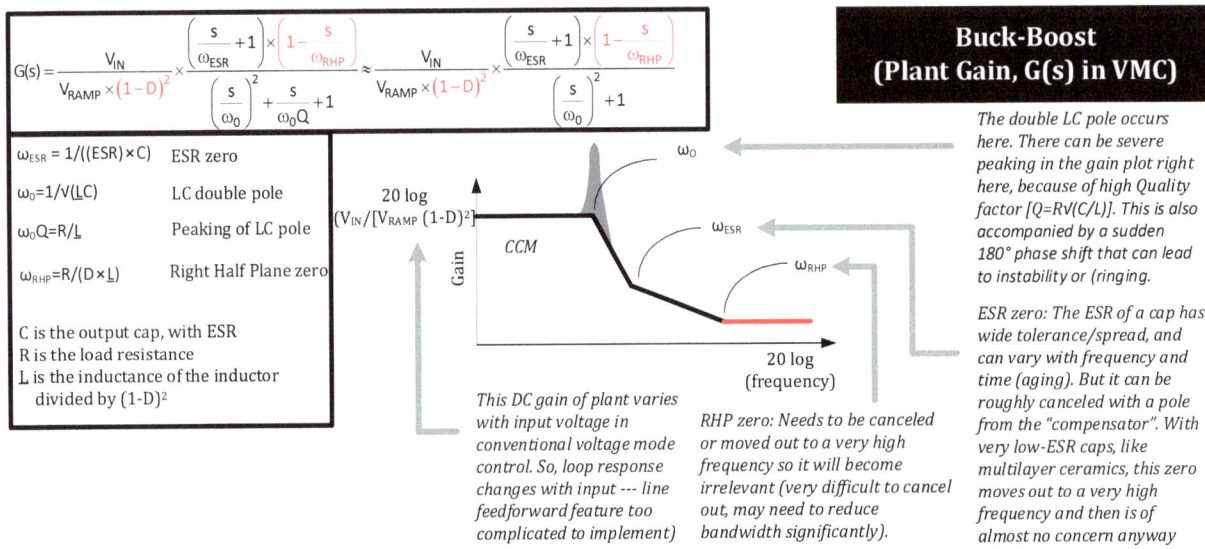

The double LC pole occurs here. There can be severe peaking in the gain plot right here, because of high Quality factor [Q=R√(C/L)]. This is also accompanied by a sudden 180° phase shift that can lead to instability or (ringing.

ESR zero: The ESR of a cap has wide tolerance/spread, and can vary with frequency and time (aging). But it can be roughly canceled with a pole from the "compensator". With very low-ESR caps, like multilayer ceramics, this zero moves out to a very high frequency and then is of almost no concern anyway

This DC gain of plant varies with input voltage in conventional voltage mode control. So, loop response changes with input --- line feedforward feature too complicated to implement)

RHP zero: Needs to be canceled or moved out to a very high frequency so it will become irrelevant (very difficult to cancel out, may need to reduce bandwidth significantly).

FIGURE 33: BUCK-BOOST IN VMC

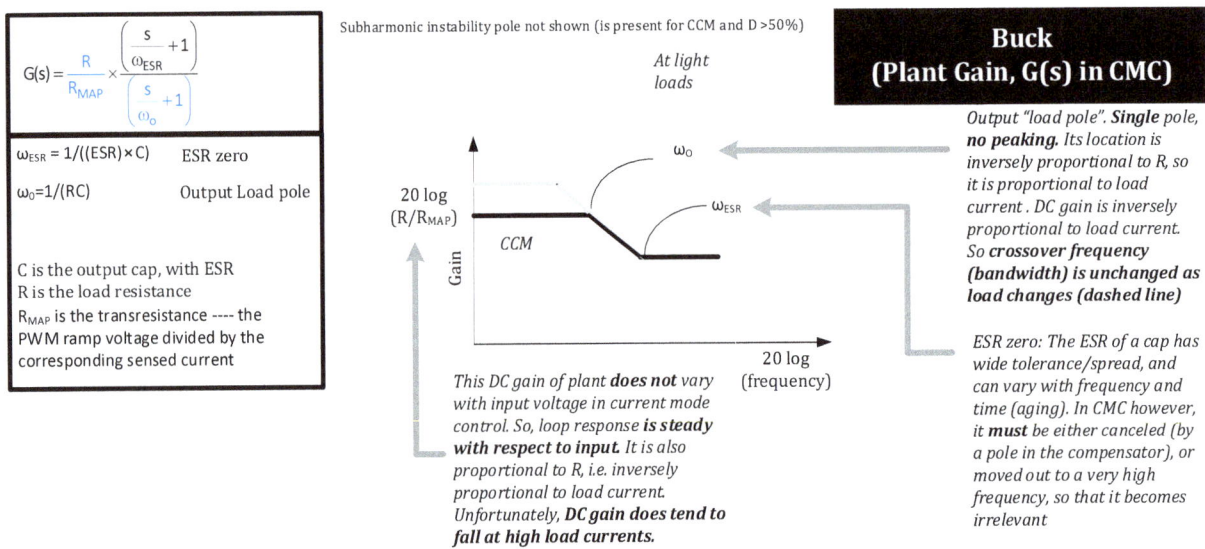

Output "load pole". **Single pole, no peaking.** Its location is inversely proportional to R, so it is proportional to load current . DC gain is inversely proportional to load current. So **crossover frequency (bandwidth) is unchanged as load changes (dashed line)**

ESR zero: The ESR of a cap has wide tolerance/spread, and can vary with frequency and time (aging). In CMC however, it **must** be either canceled (by a pole in the compensator), or moved out to a very high frequency, so that it becomes irrelevant

This DC gain of plant **does not** vary with input voltage in current mode control. So, loop response **is steady with respect to input**. It is also proportional to R, i.e. inversely proportional to load current. Unfortunately, **DC gain does tend to fall at high load currents.**

FIGURE 34: BUCK IN CMC

A simple way of collecting a Bode plot on a switcher is shown in **Figure 35**. A current loop and a *passive* current probe (snap-on coil) are the basic requirements. A standard HP/Agilent network analyzer such as the 4396B are required. No need for complicated setup from Ridley or Venable.

And indeed, as the math in **Figure 36** shows, we do measure T ("open loop gain") on the closed loop system, provided we inject the signal at a suitable point, as in **Figure 35**.

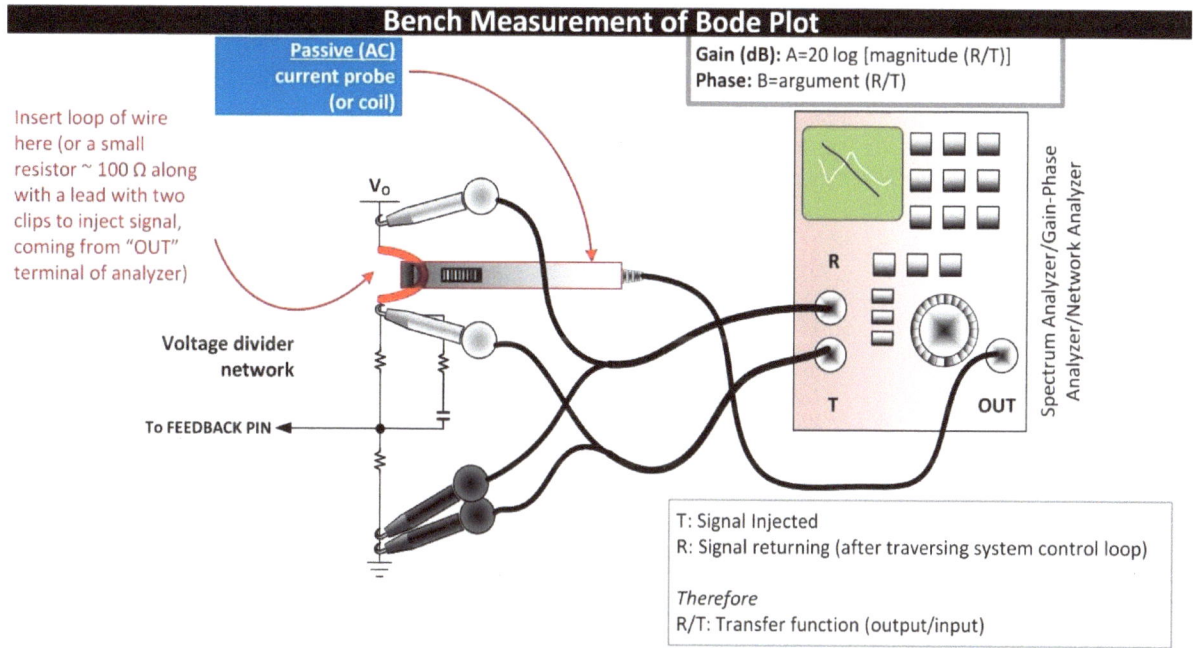

FIGURE 35: A SIMPLE WAY OF DOING A LOOP GAIN-PHASE (BODE) PLOT ON THE BENCH

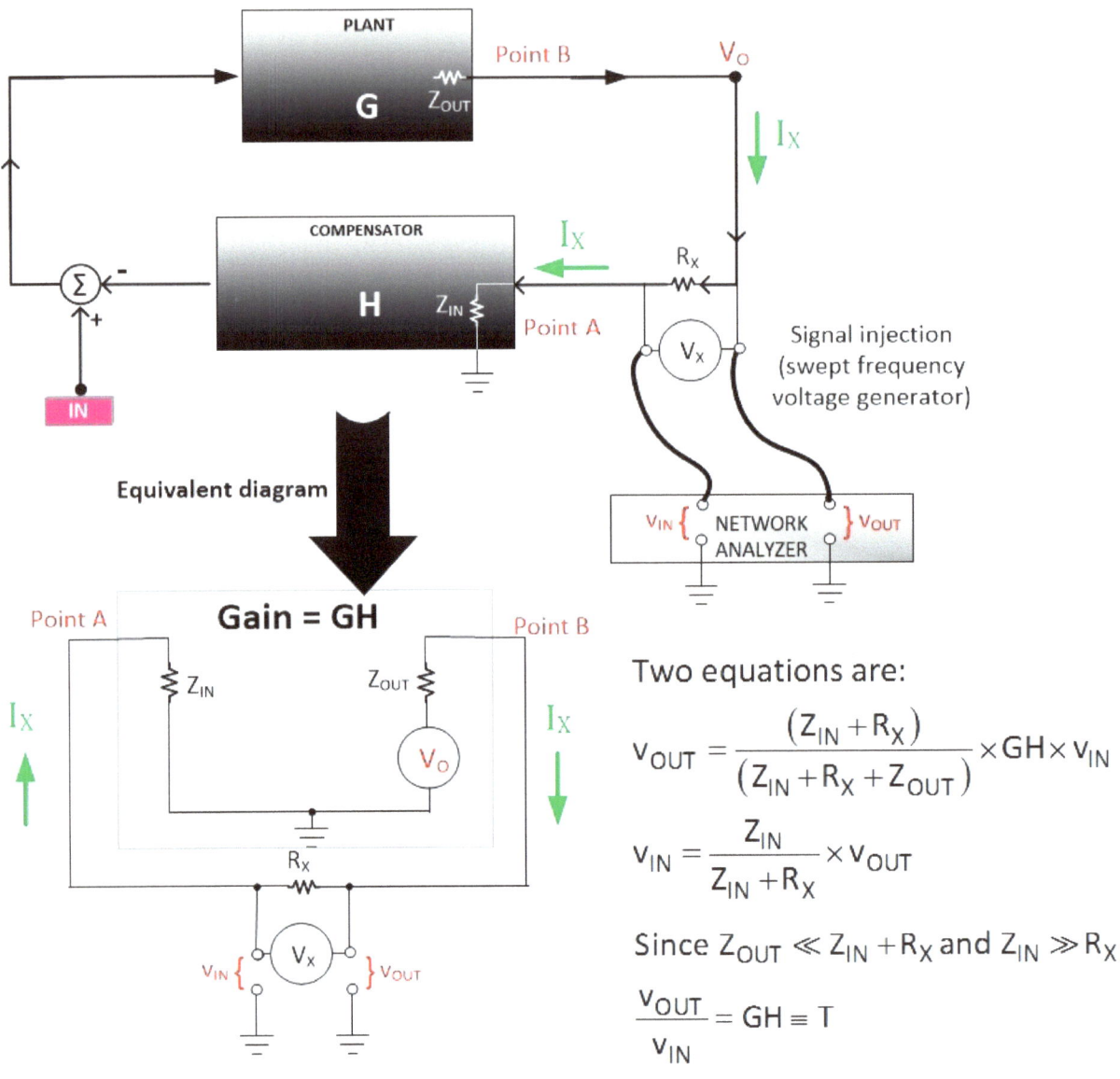

Two equations are:

$$v_{OUT} = \frac{(Z_{IN} + R_X)}{(Z_{IN} + R_X + Z_{OUT})} \times GH \times v_{IN}$$

$$v_{IN} = \frac{Z_{IN}}{Z_{IN} + R_X} \times v_{OUT}$$

Since $Z_{OUT} \ll Z_{IN} + R_X$ and $Z_{IN} \gg R_X$

$$\frac{v_{OUT}}{v_{IN}} = GH \equiv T$$

FIGURE 36: INDEED, WE MEASURE (OPEN) LOOP GAIN ON A CLOSED LOOP SETUP

One of the big nuisances regarding typical analog compensators is the difficulty of tweaking any aspect of the gain profile.

We will illustrate this with an actual example shortly. But the baseline for that is the following solved example.

Example: *Using a 300 kHz synchronous buck controller we wish to step-down 15V to 1V. The load resistor is 0.2Ω (5A). The PWM ramp is 2.14V as per the datasheet of the part. The selected inductor is 5 μH, and the output capacitor is 330 μF, with an ESR of 48 mΩ.*

We know that the plant gain at DC for a buck is $V_{IN}/V_{RAMP} = 7.009$. Therefore, $(20 \times \log)$ of this gives us 16.9 dB. The LC double pole is at

$$f_{LC} = \frac{1}{2\pi \times \sqrt{LC}} = \frac{1}{2\pi \times \sqrt{5 \times 10^{-6} \times 330 \times 10^{-6}}} \Rightarrow 3.918\,\text{kHz}$$

Here we want to set the crossover frequency of the open-loop gain at $1/6^{th}$ the switching frequency, i.e. at 50 kHz. Therefore we can solve for the integrator's fp0 and thereby its "RC" using

$$fp0 = \frac{V_{RAMP}}{V_{IN}} \times fcross \equiv \frac{1}{2\pi \times RC}$$

So in our case, the integrator's RC is

$$R_1 C_1 = \frac{V_{IN}}{2\pi \times V_{RAMP} \times fcross} = \frac{15}{2\pi \times 2.14 \times 50 \times 10^3} = 2.231 \times 10^{-5}\,\text{s}^{-1}$$

If we have selected R_1 (upper resistor of divider) as say 2kΩ, C_1 is then

$$C_1 = \frac{2.231 \times 10^{-5}}{2 \times 10^3} \Rightarrow 11.16 \text{ nF}$$

The crossover frequency of the integrator section of the op-amp is

$$fp0 = \frac{1}{2\pi \times R_1 C_1} = \frac{10^5}{2\pi \times 2.231} \Rightarrow 7.133 \text{ kHz}$$

The ESR-zero is at

$$fesr = \frac{1}{2\pi \times 48 \times 10^{-3} \times 330 \times 10^{-6}} \Rightarrow 10.05 \text{ kHz}$$

The required placement of zeros and poles is

$fz1 = fz2 = 3.918 \text{ kHz}$ (place both zeros at at the LC pole location)

$fp1 = fesr = 10.05 \text{ kHz}$ (place first pole to cancel ESR zero) s

$fp2 = 10 \times fcross = 500 \text{ kHz}$ (unoptimized/standard solution)

(We can set fp2 = f_{CROSS} for "better" phase margin)

The components required to make this happen are (solutions of several simultaneous equations)

$$C_2 = \frac{1}{2\pi \times R_1}\left(\frac{1}{fz1} - \frac{1}{fp1}\right) = \frac{1}{2\pi \times 2 \times 10^6}\left(\frac{1}{3.918} - \frac{1}{10.05}\right) \Rightarrow 12.4 \text{ nF}$$

$$R_2 = R_1 \frac{fp0}{fz2} = 2 \times 10^3 \times \frac{7.133}{3.918} \Rightarrow 3.641 \text{ k}\Omega$$

$$C_3 = \frac{1}{2\pi \times (R_2 fp2 - R_1 fp0)} = \frac{10^{-6}}{2\pi \times (3.641 \times 500 - 2 \times 7.133)} \Rightarrow 88.11 \text{ pF}$$

$$R_3 = \frac{R_1 \times fz1}{fp1 - fz1} = \frac{2 \times 10^3 \times 3.918}{10.05 - 3.918} \Rightarrow 1.278 \ k\Omega$$

We already know C_1 is 11.16nF and R_1 was selected to be 2 kΩ. So here is a summary of all the components (with the voltage divider component highlighted, to indicate it is an input):

C_1=11.16 nF, C_2= 12.4 nF, C_3= 88.11 pF, <mark>R_1= 2 k</mark>, R_2 = 3.641 k, R_3 = 1.278 k.

This baseline corresponds to the central (solid red) gain curve in **Figure 37.**

We first ask: how do we lower f_{CROSS}, *only*, without changing the basic location of the poles and zeros. In other words we simply want to translate the red solid curve vertically down. The first step is to double C_1, because R_1 and C_1 determine fp0 (the crossover of the pole-at-origin "p0" as per **Figure 38**), and R_1 is preferably fixed since it is part of the voltage divider. However, now looking at the interaction matrix in **Figure 38**, we see that C_1 is also part of the second zero "z2". And this doubling of C_1 will no doubt lower fz2. We can see this step #1, the red dashed gain curve in **Figure 37**. But that is not what we wanted. So looking again at **Figure 38**, we realize that to get fz2 back to where it was, we need to go through step #2: halve R_2.This is the blue dashed line in **Figure 37**. Unfortunately, since R_2 was also part of p2 as per **Figure 38**, halving R_2 has shifted fp2 to a higher frequency. We need to correct that too. This is done through step #3, where we double C_3. This gives us the solid blue line in **Figure 37**, and since C_3 is only part of p2, *the domino effect stops right here*, luckily.

Similarly, if we want to raise f_{CROSS}, we can go through the three steps #A, #B and #C shown in **Figure 37**.

We can achieve our target of raising or lowering f$_{CROSS}$ without changing the locations of the other poles/zeros, but with a total of three component changes!

It is not as simple as putting one decade box somewhere in the compensator, and blindly tweaking the Bode plot.

Now suppose we want to shift both coincident zeros to half their original frequency, perhaps because we changed the inductor or/and output capacitor to shift the LC pole to half the frequency. Looking at **Figure 39** we see that though shifting fz2 seems easy, we are unable to intuitively change fz1, since we get trapped in a strange circle.

Keep in mind that from the locations of the zeros, there is a constraining relationship which we may not have explicitly recognized so far.

$$fz1 = \frac{1}{2\pi(R_1 + R_3)C_2}$$

$$fz2 = \frac{1}{2\pi R_2 C_1}$$

So if fz1=fz2, we have

$$\frac{C_1}{C_2} = \frac{R_1 + R_3}{R_2}$$

This is the constraining relationship inherent in our strategy. Indeed we can confirm by plugging in the numerical values from the book, that this is true. *Further, it needs to be maintained wherever our LC pole is positioned, as per our compensation strategy.* If we do not, our simple strategy will break down, and all bets are off. We may be able to manually tweak crossover frequency and/or phase margin on the bench by using a decade box for one of the resistors involved, as is often done, but at best that would be a minor tweak. In reality as we see below, *many components have to be changed simultaneously.*

As mentioned, halving fz2 is relatively easy. All we need to do, is to double R_2. But since pole "p2" also depends on R_2, to keep it from moving we halve C_3. So that is over, because C_3 is involved only in "p2", not in any other pole or zero location.

Shifting fz1 is however very tricky, and cannot be done intuitively. First, it involves three components: R_1, R_3 and C_2. We don't want to be forced to change R_1, since that is part of the voltage divider. However, if we simply double C_2, this also affects pole "p1" and to keep that unchanged, we need to halve R_3. But R_3 is also involved in the zero "z1", and so the entire process seems convoluted. Luckily, since C_2 also gets multiplied by C_1 in the location of "z1", we do manage to move the location of z1, but by a certain weighted amount, based on the value of R_3.

$$fz1 = \frac{1}{2\pi(R_1 + R_3)C_2}$$

In other words, it is hard to predict what the values of the RC's are for changing fz1. We need to go back to the basic equations for calculating all the components from scratch (mathematics, not intuition). In our specific numerical example, we recalculate all the values if we change fLC (fz1 and fz2) from 3.918 kHz to 3.918/2 = 1.959 kHz. The before and after RC values are:

Before:

C_1=11.16 nF, C_2= 12.4 nF, C_3= 88.11 pF, R_1= 2 k, R_2 = 3.641 k, R_3 = 1.278 k.

After:

C_1=11.16 nF, C_2= 32.7 nF, C_3= 44 pF, R_1= 2 k, R_2 = 7.28 k, R_3 = 484.24.

Our conclusion is, just to shift the two zeros without changing the crossover frequency and the other poles and zeros of a Type 3 compensator, we need *four* component values to be changed every time, and it isn't straightforward either. We certainly can't do it "on the fly". But we can do that with digital techniques, as we will learn in the next part of this series.

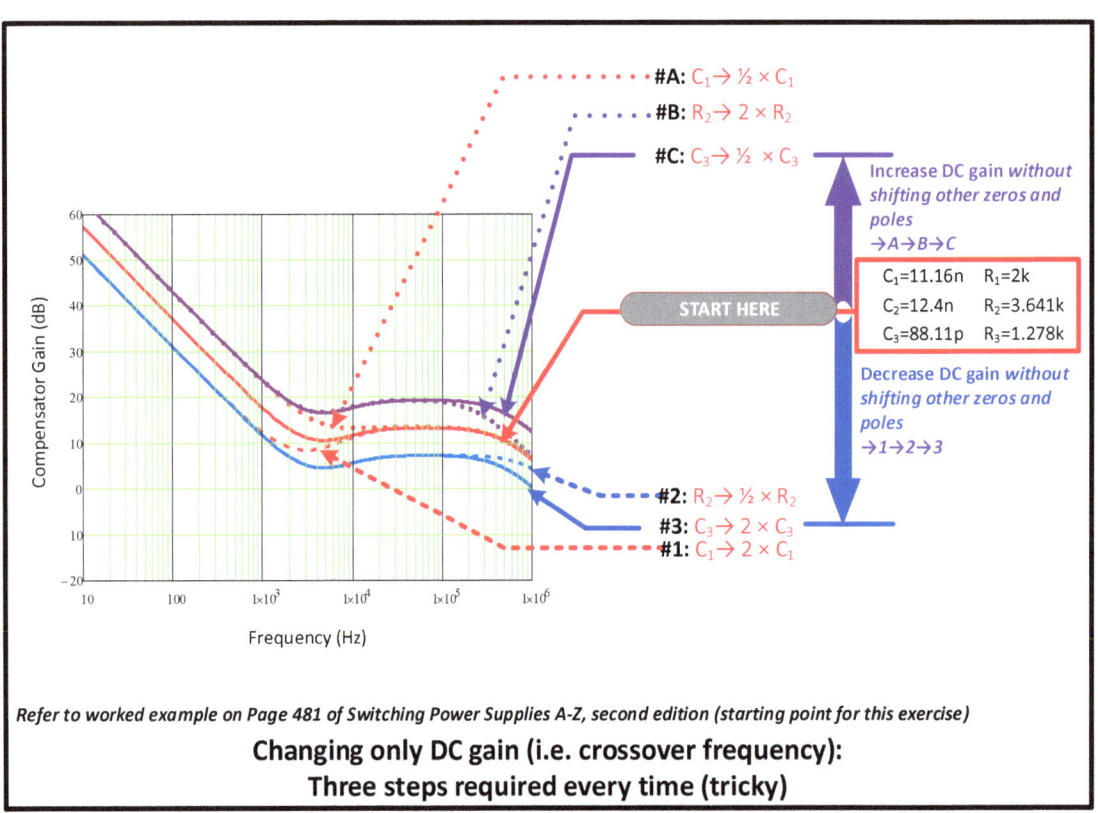

FIGURE 37: CHANGING CROSSOVER FREQUENCY (ONLY) USING A TYPE 3 COMPENSATOR

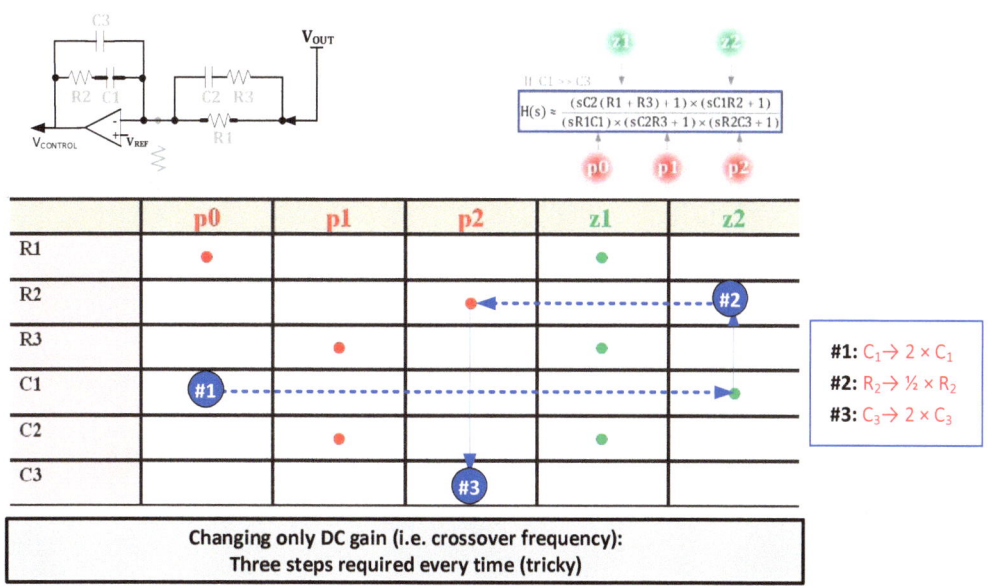

FIGURE 38: HOW WE ADJUST COMPONENT VALUES FOR PREVIOUS FIGURE

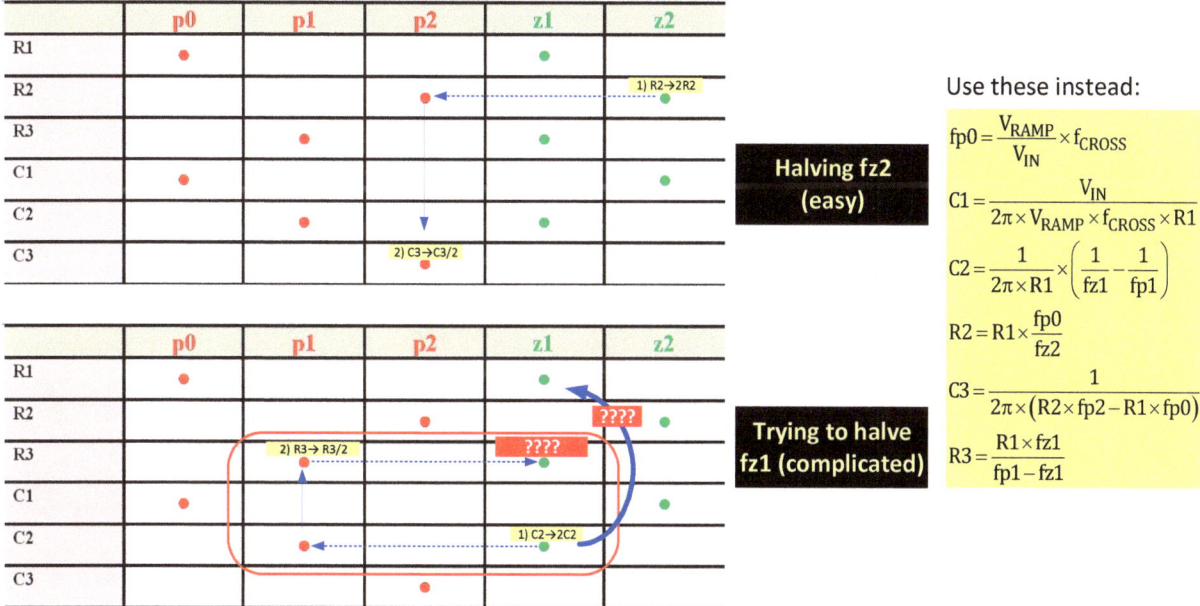

FIGURE 39: TRYING TO INTUITIVELY SHIFT BOTH COINCIDENT ZEROS (ONLY)

Unfortunately, the trouble doesn't stop with our inability to tweak a Type 3 compensator easily or intuitively. For if we look at the capacitor values we have calculated for our application, they are not even close to standard values. Capacitors still come mainly in the E12 series: 10 12 15 18 22 27 33 39 47 56 68 82. The tolerance is +,- 10%. On top of that, unless we are using C0G capacitors, we have to include the effects of temperature, voltage, aging and so on. *Not to forget that our initial equations were based on an assumption: $C_1 >> C_3$.*

So the final placement of the poles and zeros, as also the bandwidth (f_{CROSS}) may be quite different from what we intended.

However just in case we want to derive more exact equations, without any approximations, here they are, with and without the $C_1 >> C_3$ approximation. But now we realize that for example, "p0", the pole at origin, is actually affected by *three* components: R_1, C_1 and C_3.

$$fp0 = \frac{1}{2\pi \times R_1(C_1 + C_3)} \approx \frac{1}{2\pi \times R_1 C_1}$$

$$fp1 = \frac{1}{2\pi \times R_3 C_2}$$

$$fp2 = \frac{1}{2\pi \times R_2 \left(\frac{C_1 C_3}{C_1 + C_3} \right)} = \frac{1}{2\pi \times R_2} \left(\frac{1}{C_1} + \frac{1}{C_3} \right) \approx \frac{1}{2\pi \times R_2 C_3}$$

$$fz1 = \frac{1}{2\pi \times (R_1 + R_3)C_2}$$

$$fz2 = \frac{1}{2\pi \times R_2 C_1}$$

Matters get even more complicated and nothing is intuitive anymore.

A POWERFUL NEW WEAPON ARRIVES ON OUR DOORSTEP

We have by now realized some of the inadequacies or weaknesses of analog compensation, among them the inability to accurately fix the loop gain shape we desire. But the most alarming weakness of Type 3 compensation is yet to come! That will be revealed in the next part. And we will show how using *digital techniques based on PID coefficients* can help us dramatically in this regard.